# CONTENTS

# COMBATING TERRORISM: PREPARING AND FUNDING FIRST RESPONDERS

# HEARING

BEFORE THE

SUBCOMMITTEE ON NATIONAL SECURITY, EMERGING THREATS AND INTERNATIONAL RELATIONS

OF THE

COMMITTEE ON GOVERNMENT REFORM

HOUSE OF REPRESENTATIVES

ONE HUNDRED EIGHTH CONGRESS

FIRST SESSION

SEPTEMBER 9, 2003

**Serial No. 108–111**

Printed for the use of the Committee on Government Reform

Available via the World Wide Web: http://www.gpo.gov/congress/house
http://www.house.gov/reform

U.S. GOVERNMENT PRINTING OFFICE

92–127 PDF WASHINGTON : 2004

For sale by the Superintendent of Documents, U.S. Government Printing Office
Internet: bookstore.gpo.gov Phone: toll free (866) 512–1800; DC area (202) 512–1800
Fax: (202) 512–2250 Mail: Stop SSOP, Washington, DC 20402–0001

## COMMITTEE ON GOVERNMENT REFORM

TOM DAVIS, Virginia, *Chairman*

DAN BURTON, Indiana
CHRISTOPHER SHAYS, Connecticut
ILEANA ROS-LEHTINEN, Florida
JOHN M. McHUGH, New York
JOHN L. MICA, Florida
MARK E. SOUDER, Indiana
STEVEN C. LATOURETTE, Ohio
DOUG OSE, California
RON LEWIS, Kentucky
JO ANN DAVIS, Virginia
TODD RUSSELL PLATTS, Pennsylvania
CHRIS CANNON, Utah
ADAM H. PUTNAM, Florida
EDWARD L. SCHROCK, Virginia
JOHN J. DUNCAN, Jr., Tennessee
JOHN SULLIVAN, Oklahoma
NATHAN DEAL, Georgia
CANDICE S. MILLER, Michigan
TIM MURPHY, Pennsylvania
MICHAEL R. TURNER, Ohio
JOHN R. CARTER, Texas
WILLIAM J. JANKLOW, South Dakota
MARSHA BLACKBURN, Tennessee

HENRY A. WAXMAN, California
TOM LANTOS, California
MAJOR R. OWENS, New York
EDOLPHUS TOWNS, New York
PAUL E. KANJORSKI, Pennsylvania
CAROLYN B. MALONEY, New York
ELIJAH E. CUMMINGS, Maryland
DENNIS J. KUCINICH, Ohio
DANNY K. DAVIS, Illinois
JOHN F. TIERNEY, Massachusetts
WM. LACY CLAY, Missouri
DIANE E. WATSON, California
STEPHEN F. LYNCH, Massachusetts
CHRIS VAN HOLLEN, Maryland
LINDA T. SANCHEZ, California
C.A. "DUTCH" RUPPERSBERGER, Maryland
ELEANOR HOLMES NORTON, District of Columbia
JIM COOPER, Tennessee
CHRIS BELL, Texas

———

BERNARD SANDERS, Vermont (Independent)

PETER SIRH, *Staff Director*
MELISSA WOJCIAK, *Deputy Staff Director*
ROB BORDEN, *Parliamentarian*
TERESA AUSTIN, *Chief Clerk*
PHILIP M. SCHILIRO, *Minority Staff Director*

### SUBCOMMITTEE ON NATIONAL SECURITY, EMERGING THREATS AND INTERNATIONAL RELATIONS

CHRISTOPHER SHAYS, Connecticut, *Chairman*

MICHAEL R. TURNER, Ohio
DAN BURTON, Indiana
STEVEN C. LATOURETTE, Ohio
RON LEWIS, Kentucky
TODD RUSSELL PLATTS, Pennsylvania
ADAM H. PUTNAM, Florida
EDWARD L. SCHROCK, Virginia
JOHN J. DUNCAN, Jr., Tennessee
TIM MURPHY, Pennsylvania
WILLIAM J. JANKLOW, South Dakota

DENNIS J. KUCINICH, Ohio
TOM LANTOS, California
BERNARD SANDERS, Vermont
STEPHEN F. LYNCH, Massachusetts
CAROLYN B. MALONEY, New York
LINDA T. SANCHEZ, California
C.A. "DUTCH" RUPPERSBERGER, Maryland
CHRIS BELL, Texas
JOHN F. TIERNEY, Massachusetts

EX OFFICIO

TOM DAVIS, Virginia
HENRY A. WAXMAN, California

LAWRENCE J. HALLORAN, *Staff Director and Counsel*
R. NICHOLAS PALARINO, *Senior Policy Advisor*
ROBERT A. BRIGGS, *Clerk*
DAVID RAPALLO, *Minority Counsel*

# COMBATING TERRORISM: PREPARING AND FUNDING FIRST RESPONDERS

---

**TUESDAY, SEPTEMBER 9, 2003**

HOUSE OF REPRESENTATIVES,
SUBCOMMITTEE ON NATIONAL SECURITY, EMERGING THREATS AND INTERNATIONAL RELATIONS,
COMMITTEE ON GOVERNMENT REFORM,
*Washington, DC.*

The subcommittee met, pursuant to notice, at 10 a.m., in room 2203, Rayburn House Office Building, Hon. Christopher Shays (chairman of the subcommittee) presiding.

Present: Representatives Shays, Turner, Murphy, Kucinich, Maloney, Dutch, and Tierney.

Staff present: Lawrence Halloran, staff director and counsel; R. Nicholas Palarino, Ph.D, senior policy advisor; Robert A. Briggs, clerk; Chris Skaluba, Presidential management intern; David Rapallo, minority counsel; Michael Yeager, minority deputy chief counsel; and Jean Gosa, minority assistant clerk.

Mr. SHAYS. A quorum being present, the Subcommittee on National Security, Emerging Threats and International Relations hearing entitled, "Combating Terrorism: Preparing and Funding First Responders" is called to order.

Well before September 11, 2001, this subcommittee heard testimony from first responders and other experts expressing frustration over the extent and pace of Federal counterterrorism equipment and training programs. They told us fragmentation and duplication hobbled a multiagency, multibillion-dollar preparedness effort that failed to answer the fundamental question, "Prepared for what?"

Since the September 11 attacks, much has been done and much more has been spent to consolidate and focus Federal support for first responders. But a growing body of analysis and commentary suggests increased activity still may not be producing enough measurable progress toward the elusive goal of actual preparedness.

One recent report of an independent task force sponsored by the Council of Foreign Relations [CFR], concluded the Nation's emergency responders remain, "dangerously ill-prepared to handle a catastrophic attack on American soil." While acknowledging some improvement in counterterrorism capabilities the CFR report says Federal funding for first responders may fall $98 billion short of meeting basic needs for training and equipment over the next 5 years.

Just as ominously, the report says any effort to quantify the cost of preparedness will be confounded by the lack of agreed-upon measures for success. "Without establishing minimal preparedness levels and equipment and performance standards that the Federal Government and State and local communities can strive to attain, the United States will have created an illusion of preparedness based on boutique funding initiatives without being systematically prepared."

These are not abstract policy considerations. Without standards, time and money will be wasted on a dangerous and costly illusion while police officers, firefighters, emergency medical teams, public health providers, and emergency managers confront terrorism without the tools they need. Nor can we afford to wait for a national consensus on standards to emerge before funding critical first responders initiatives. The threat is now, the threat is real, and local emergency responders need to be prepared to meet it.

Next Monday the city of Stanford, CT, will conduct a regional table-top exercise of emergency response capabilities against a weapon of mass destruction scenario sponsored by the Department of Homeland Security. The simulation will help emergency responders to assess and improve readiness against terrorist attacks. To sharpen our focus on the needs of our first responders, the subcommittee will be there to observe the exercises and take testimony from participants on the scope and impact of Federal efforts to enhance local preparedness.

This morning, we are joined by two distinguished expert panels of witnesses. We appreciate their time, talent and dedication, and we are very grateful for their continued contributions to our oversight work.

At this time the Chair will recognize the ranking member, Mr. Kucinich.

[The prepared statement of Hon. Christopher Shays follows:]

ONE HUNDRED EIGHTH CONGRESS

# Congress of the United States

## House of Representatives

COMMITTEE ON GOVERNMENT REFORM

2157 RAYBURN HOUSE OFFICE BUILDING

WASHINGTON, DC 20515–6143

MAJORITY (202) 225-5074
FACSIMILE (202) 225-3974
MINORITY (202) 225-5051
TTY (202) 225-6852

www.house.gov/reform



SUBCOMMITTEE ON NATIONAL SECURITY, EMERGING THREATS, AND INTERNATIONAL RELATIONS

Christopher Shays, Connecticut
Chairman
Room B-372 Rayburn Building
Washington, D.C. 20515
Tel: 202 225-2548
Fax: 202 225-2382
E-mail: hr.groc@mail.house.gov

# Statement of Rep. Christopher Shays
## September 9, 2003

Well before September 11, 2001, this Subcommittee heard testimony from first responders and other experts expressing frustration over the extent and pace of federal counterterrorism equipment and training programs. They told us fragmentation and duplication hobbled a multi-agency, multi-billion dollar preparedness effort that failed to answer the fundamental question, "Prepared for what?"

Since the 9/11 attacks, much has been done, and much more has been spent, to consolidate and focus federal support for first responders. But a growing body of analysis and commentary suggests increased activity still may not be producing enough measurable progress toward the elusive goal of actual preparedness.

One recent report of an Independent Task Force sponsored by the Council of Foreign Relations (CFR) concluded the nation's emergency responders remain "dangerously ill-prepared to handle a catastrophic attack on American soil." While acknowledging some improvement in counterterrorism capabilities, the CFR report says federal funding for first responders may fall 98 billion dollars short of meeting basic needs for training and equipment over the next five years.

Just as ominously, the report says any effort to quantify the cost of preparedness will be confounded by the lack of agreed-upon measures for success: "Without establishing minimal preparedness levels and equipment and performance standards that the federal government and state and local communities can strive to attain, the United States will have created an illusion of preparedness based on boutique funding initiatives without being systematically prepared."

These are not abstract policy considerations. Without standards, time and money will be wasted on a dangerous and costly illusion, while police officers, firefighters, emergency medical teams, public health providers and emergency managers confront terrorism without the tools they need. Nor can we afford to wait for a national consensus on standards to emerge before funding critical first responder initiatives. The threat is now, the threat is real, and local emergency responders need to be prepared to meet it.

Next Monday, the City of Stamford, Connecticut will conduct a tabletop exercise of emergency response capabilities against a weapon of mass destruction scenario. Sponsored by the Department of Homeland Security, the simulation will help emergency responders there assess and improve readiness against terrorist attacks. To sharpen our focus on the needs of first responders, the Subcommittee will be there to observe the exercise and take testimony from participants on the scope and impact of federal efforts to enhance local preparedness.

This morning, we are joined by two distinguished expert panels of witnesses. We appreciate their time, talent and dedication and we are grateful for their continued contributions to our oversight work.

Mr. KUCINICH. From a Midwest region, Mr. Chairman.

Thank you very much, Mr. Chairman, for your ongoing efforts on this very important concern that all Americans have about preparing and funding first responders. And I want to thank the witnesses. Senator Rudman, thank you for your consistent leadership on these matters and, Dr. Smithson, for your participation and your testimony.

And I welcome this hearing today on preparing and funding first responders. As our witnesses today will point out, the fire, police, and emergency medical personnel who need to respond to catastrophic terrorist attacks are underfunded and they are unprepared for a wide range of threats. Firefighters around the country only have half the radios they need to equip a shift. Two-thirds don't have the breathing apparatus they need to work in hazardous conditions. Police departments don't have protective gear to secure a site following an attack with weapons of mass destruction. Public health officials are overwhelmed by requests for tests but still don't have the basic equipment to identify chemical or biological agents.

The Council on Foreign Relations released a report earlier this summer concluding that we are spending as little as one-third of what we need to develop adequate first responder capabilities. To meet this need, the Federal Government would have to increase its current spending, $5.4 billion, five times to an annual contribution of $25 billion.

This past Sunday the President told the American people, quote, "This will take time and require sacrifice." He said without a trace of irony, "We will do what is necessary, we will spend what is necessary to achieve this essential victory in the war on terror."

From the President's words, you would think he was talking about giving firefighters and EMTs the basic equipment they need to do their jobs and that could save their lives in a terrorist attack. You would think he would request $87 billion additional to meet this need. But the focus of his attention in Iraq to pay for the war that we didn't need and that didn't buy us one iota of security from terror has created greater concerns among the American people. The President and his administration were not forthcoming with the American people by suggesting that the war in Iraq is part of the war on terror. Iraq had nothing do with September 11. There was no connection between Saddam Hussein and al Qaeda. In fact, the war on Iraq has produced the opposite effect and created a vacuum where al Qaeda, which was not operating inside Iraq before the war, is now targeting our men and women.

Mr. Chairman, much of the report that prompted this hearing calls on Congress to increase funding for first responders, require minimum national standards for preparedness, and create a sensible system to allocate resources. Congress should step up to the plate. It should do it, first, by rejecting the President's request for funding of the war, and should instead cause us to bring our troops home. For the safety and security of our troops, I say bring them home. And for the safety of our first responders, let's start funding police, fire, and EMT to levels that they need to protect this country. Thank you.

Mr. SHAYS. Thank the gentleman.

[The prepared statement of Hon. Dennis J. Kucinich follows:]

**Statement of Rep. Dennis Kucinich**
**Hearing on "Combating Terrorism:**
**Preparing and Funding First Responders"**
**September 9, 2003**

Mr. Chairman, I welcome this hearing today on preparing and funding first responders.

As our witnesses today will point out, the fire, police, and emergency medical personnel who need to respond to catastrophic terrorist attacks are underfunded and unprepared for a wide range of threats. Firefighters around the country only have half the radios they need to equip a shift. Two-thirds don't have the breathing apparatus they need to work in hazardous conditions. Police Departments don't have protective gear to secure a site following an attack with weapons of mass destruction. Public health officials are overwhelmed by requests for tests but still don't have basic equipment to identify chemical or biological agents.

The Council on Foreign Relations released a report earlier this summer concluding that we are spending as little as one-third of what we need to develop adequate first responder capabilities. To meet this need, the federal government would have to increase it current spending of $5.4 billion five times to an annual contribution of $25 billion.

This past Sunday, the President told the American people that "This will take time and require sacrifice." He said, without a trace of irony, "We will do what is necessary, we will spend what is necessary, to achieve this essential victory in the war on terror." From the President's words, you would think he was talking about giving firefighters and EMTs the basic equipment they need to do their jobs, that could save their lives, in a terrorist attack.

You would think he would request $87 billion additional dollars to meet this need. But the focus of his attention is Iraq, to pay for a war that we didn't need and that didn't buy us one iota of security from terror.

Mr. Chairman, much of the report that prompted this hearing calls on Congress to increase funding for first responders, require minimum national standards for preparedness, and create a sensible system to allocate resources. Congress should step up to the plate. It should do it first by rejecting the President's request for Iraq reconstruction and focusing it where it belongs – on our needs at home.

Mr. SHAYS. At this time the Chair would recognize the vice chairman of the committee, Mr. Turner.

Mr. TURNER. Mr. Chairman, thank you for your continued leadership in focusing on the issue of preparedness of our country in responding to terrorist attacks and for holding this important hearing on the funding and policy priorities for our Nation's first responders. This is an appropriate time for us to examine the needs of our first responders during the week of the second anniversary of terrorist attacks on the World Trade Center and the Pentagon and after the recently published report, "Emergency Responders: Drastically Underfunded, Dangerously Unprepared."

I want to thank our witnesses for being here today. They are certainly preeminent experts on homeland security and first responders. As a former mayor of Dayton, OH, I am especially interested in examining the Federal role in aiding first responders. Dayton, OH was one of the few cities that had a weapons of mass destruction preparedness exercise prior to September 11, exercises that Attorney General John Ashcroft had attended.

Federal assistance to local first responders became even more crucial and valuable since the attacks of September 11, 2001. Police officers, firefighters, and EMTs are being asked to train and prepare for events that are even more terrifying and destructive than the acts of crime and violence that they see on a routine basis. It was local first responders who bravely answered the call on September 11, and it will be the local teams who respond to future attacks. And I am pleased that we will hear from two fire chiefs whose departments responded to the Pentagon attack on our second panel today.

The Council on Foreign Relations report, "Emergency First Responders: Drastically Underfunded, Dangerously Unprepared" identifies many areas where the Federal, State, and local governments need to increase their funding, improve their cooperation and communication, and invest in training and equipment.

The threat of terrorism is unknown, incredibly difficult to prepare for. This report is the first of its kind, an attempt to account for across-the-board first responder spending. I look forward to hearing how Congress can effectively assist first responders and how the Federal Government can efficiently maximize the dollars it spends on first responders and homeland security. I am particularly interested in the issue of how the Federal Government can assist in coordination and benchmarking the needs of individuals and our people who are on the front line, our first responders. Thank you.

Mr. SHAYS. At this time the Chair would thank the gentleman and recognize the distinguished lady from New York, Mrs. Maloney, who just recently returned from Iraq.

Mrs. MALONEY. Thank you, Mr. Chairman.

Yes, with your leadership there on the issues that we were confronting in another problem area. And I thank very much the Honorable Senator Rudman and Dr. Smithson for joining us today to discuss the critical issue. And as chair of the Democratic Caucus's Task Force on Homeland Security and as a Member who represents Manhattan and Queens and one who lost, I should say, 500 constituents on September 11, you are obviously talking about an issue that is very close to our hearts. In fact, the safety and security of our homeland is something we can all agree on, and it is truly a bipartisan issue because we are all advocates of tough homeland security and we should all celebrate the successes and work together to fix the vulnerabilities.

I truly applaud the Council on Foreign Relations for working in a bipartisan vein to help all of us better understand the holes that must be patched and for really changing the debate about homeland security. Your report has had a tremendous impact on the thinking of Members of Congress.

I have met with Jamie Metzl, the members of my task force, and in fact he addressed the entire Democratic Caucus on the findings of the report and the importance of it. And we thank you. Many people were awed by the report's estimate that we must spend almost $100 billion more on emergency responders. It is a very important report that really hit the mark and we do thank you for it.

My home city is New York and we are target No. 1. When the rest of the country is orange, we are red. And today we are just 2 days before the second anniversary of September 11. And this is the news from New York: Just last week the city received its very first Federal homeland security grant of $30 million. That's $30 million to prepare New York City for another attack and $87 billion to rebuild Iraq. Our mayor tells us that we spend $23 million a week just making sure that New York is safe, and we applaud him and the police and responders for having thwarted some attempted attacks to New York.

And here's another story that's almost unbelievable. The firefighters in New York City, they told me that they are less prepared today than they were before September 11. They say that the famous communication radios that didn't work on September 11 still don't work. The city has closed six firehouses this year. I'm gagging when I give you this information. And there are 530 fewer firefighters on the street at any given time in New York City today than there were on September 10, 2001. That is hard to believe, but those are the facts given to me yesterday from the New York City Fire Department.

Our task force has been learning more about the needs in hometowns across the country from the men and women on the front lines. They say that homeland security is a strain on local governments; that there has been a lack of guidance from the Department of Homeland Security and that there has been a lack of money from the Federal Government. It is one thing when people in Washington are talking about homeland security needs. It is another when the men and women dedicating their lives to making ours safe from terrorism say over and over again to anyone who will listen that they need more help.

So, I am very very thankful that you are here today. I look forward to hearing your comments. I have read your report many times and I am deeply grateful for the tremendous leadership role that you have played in focusing debate in a bipartisan way on homeland security needs. You have not gotten enough appreciation from the public or from Congress for your hard work and so I thank you for having this hearing today. It's important.

[The prepared statement of Hon. Carolyn B. Maloney follows:]

*14th District • New York*

Congresswoman
*Carolyn* **Maloney**
**Reports**

2331 Rayburn Building • Washington, DC 20515 • 202-225-7944
1651 Third Avenue • Suite 311 • New York, NY 10128 • 212-860-0606
www.house.gov/maloney/press.html

Opening Statement
Rep. Carolyn Maloney (NY-14)
Gov. Reform National Security Subcommittee
Hearing on Council on Foreign Relations
Emergency Responders report
September 9, 2003

Thank you, Mr. Chairman, and thank you Senator Rudman, Mr. Clarke and Dr. Smithson for joining us today to discuss this critical issue.

As Chair of our caucus's Task Force on Homeland Security and as a Member who represents parts of Manhattan and Queens, homeland security and emergency responders are issues I care deeply about.

In fact, the safety and security of our homeland is something we can all agree on – it is truly a bipartisan issue. Because we are all advocates of tough homeland security, we should all celebrate our successes and work together to fix vulnerabilities.

I applaud the Council on Foreign Relations for working in this bipartisan vein to help us better understand the holes that must be patched and for really changing the debate about hometown security.

By bringing together respected, thoughtful leaders from both sides of the aisle, two of whom are sitting before us, the Council produced a report that has garnered everyone's attention. My Task Force has used your report extensively and have met with the Jamie Metzl, who led the Council's efforts.

Many people were awed by the report's estimate that we must spend almost $100 billion more on emergency responders. It is a report that most of us really took to heart.

Senator Rudman and Mr. Clarke, I truly believe that the Administration badly served your efforts with its reaction to the report, saying that the additional $100 billion for would just buy "gold-plated telephones." While everyone else was treating this as serious business, the Department of Homeland Security was treating it as a joke.

You were in high demand to appear at Congressional hearings, but from what I understand, the Department of Homeland Security didn't even try to arrange a meeting. What in the world are they doing over there?

In case they still think that the money would be a waste, here's a tidbit from my hometown, the number one terrorist target, New York City.

Today, we are just 2 days before the second anniversary of 9/11, and this is the news: just last week, the city received its very first federal homeland security grant of $30 million. That's $30 million to prepare New York for another attack and $87 billion to rebuild Iraq.

Here's another sad story: the firefighters in New York City told me that they are less prepared today than they were before 9/11. They say the communications radios that didn't work on 9/11 still don't work today. The city closed six firehouses this year. There are 530 *fewer* firefighters on the street at any given time in New York City today than there were on September 10, 2001.

Can you believe that?

Our Task Force has been learning more the needs in hometowns across the country from the men and women on the front lines. They say that homeland security is a strain on local governments, that there has been a lack of guidance from the Department of Homeland Security, and that there has been a lack of money from the federal government.

It's one thing when people in Washington are talking about homeland security needs. It's another when the men and women dedicating their lives to making ours safe from terrorism say over and over again to anyone who will listen that they need more help.

I'm glad we're here today to listen to and discuss what emergency responders said to the Council.

Thank you for your hard work, and I look forward to your testimony.

Mr. SHAYS. You don't think he's gotten enough appreciation?

Mrs. MALONEY. No, I don't. I don't, because it certainly hasn't been translated into a direction of policy or dollars or the bottom line of taking the recommendations and turning them into action.

Mr. SHAYS. Let the day begin here.

Mrs. MALONEY. OK.

Mr. SHAYS. Mr. Ruppersberger.

Mr. RUPPERSBERGER. OK. Mr. Chairman, first thing, I want to really applaud the Council on Foreign Relations for the bipartisan approach to deal with this issue. During the September 11 incident, I was a county executive and understand the needs of putting your first responders, your police, your firefighters, your paramedics, and spending the overtime and the money that is necessary. But what really a lot of this comes down to in my opinion, and I hope that we can develop this issue today, is the matter of priority of funding.

There's no question that once our Commander in Chief, the President, decided to go to Iraq that we had to stand behind, and I think that we have to do what we have to do—and we won't get into that issue now—with respect to our military and give them the resources that they need. But our first responders need the resources also.

In the Second Congressional District of Maryland that I represent, we have NSA, we have two Army bases—Fort Meade and Aberdeen. I have BWI Airport and the whole port of Baltimore. And there are a lot of issues involving the resources that are needed to detect any weapons of mass destruction, but also to deal with issues involving the intelligence that is needed, working with the State, Federal, and local authorities.

So from my perspective from being in local government and now coming into the Federal Government, it's a matter of priorities. The tax cuts just aren't working. We need to reevaluate our entire situation now. We have never in the history of our country been at war—and we are at war with Iraq, we are at war with Afghanistan, and we are at war with terrorism. We must reprioritize and give the resources to first responders. Thank you.

Mr. SHAYS. I thank the gentleman.

And we have just been joined by Mr. Murphy. Mr. Murphy, do you have any statement before we begin?

Mr. MURPHY. Not yet. Thank you.

Mr. SHAYS. OK. Thanks for being here.

And Mr. Tierney. I'm sorry, the Chair would recognize Mr. Tierney.

Mr. TIERNEY. Good morning and thank you for being here with us this morning. I just have some brief remarks about local preparedness. I have just had the opportunity to meet with most of my chiefs of fire and police and other first responders and to do a bit of a survey on how they felt the Federal Government was dealing with the situation. And I think it's pretty clear that they have done what they have been asked to do. You know, they have planned—immediately after September 11 when we gathered everybody together, they were told to put their plans together, those that hadn't already done that, and they did put their plan together. They have expended tremendous amounts of resources on overtime, personnel and training, and what equipment they could gather in some pretty difficult times and climate in their own States and localities.

They could certainly use, they tell me, some minimum national standards of just what it is that they might be expected to do in particular situations, and they look forward to some guidance on that end of it. Particularly I think one of my chiefs mentioned to me that when it goes from yellow to orange, what? You know, just what are they supposed to do? And what is enough security at a nuclear power plant or what is enough security at a chemical manufacturing plant? And just some minimum guidelines there would be extremely helpful to them.

They need a better communication system, obviously. And I think that's something that we have had a great deal of conversation with. I know that there are plans in the works now to put an interoperable system together. We have had some folks out of Cambridge, Professor John Donovan and others, working on that. And they have a prototype going on the military side of things. We'd like to see one done in a couple of districts so that we could get an idea of how it works and then bring it up to scope and scale so that everybody is on the same page when an incident occurs as to what is going on and how people should react and be able to do that.

So they need the resources. They would prefer it without State bureaucracy getting between the Federal distribution and them. They feel that they're really constrained on flexibility. If it comes down and they're not allowed to use it on overtime, sometimes it isn't of as much valuable to them. Or if they need equipment and it's only limited to personnel, those type of issues still remain unresolved. So all of those things have come to their attention.

They have needs of overtime, personnel constraints, communication, and some guidance. And I hope that in the context of your testimony this morning, you will touch on those points and let us get back to them with how we might be doing in that area and move together toward resolving these issues.

I agree with my colleague here that we are spending enormous sums of money to secure Iraq and to give people there a sense of security. I think we should have put a comparable amount of attention to making the people in this country feel secure as they move forward, and I think we can do that and I look forward to your cooperation in doing that. Thank you. Thank you.

Mr. SHAYS. I thank the gentleman. Before recognizing the panel, let me just deal with two housekeeping measures. I ask unanimous consent that all members of the subcommittee be permitted to place an opening statement in the record, and the record will remain open for 2 days for that purpose. And, without objection, so ordered. I ask further unanimous consent that all witnesses be permitted to include their written statements in the record. Without objection, so ordered.

We are very fortunate and I think you're seeing this reflected in the participation of the membership, both on the issue and in terms of our participants on our two panels. We are fortunate to have Senator Warren Rudman who is chairman of the Independent Task Force on Emergency Responders, Council on Foreign Relations. This is the genesis of what this hearing is about. He's currently a partner at Paul, Weiss, Rifkind, Wharton and Garrison. And he's a two-term Senator from New Hampshire and this committee has known him in particular for his outstanding work on the Hart-Rudman Commission. As many will remember, there were three commissions, the Bremer Commission, the Gillmore Commission and the Hart-Rudman Commission. They all agreed on understanding what the terrorist threat was, developing a strategy to respond to it, and to reorganize to implement that strategy. But the Hart-Rudman Commission believed we needed the most radical reorganization, and ultimately that's what this Congress decided to do with the Department of Homeland Security. Great to have you here, Senator Rudman.

We are also joined by Dr. Amy Smith, senior fellow, Center for Strategic and International Studies as of last week, and previous senior associate at the Simpson Center's directed chemical and biological weapons nonproliferation project.

If I could get you to stand up. As you know, this committee swears in our witnesses and then we'll hear your testimony.

[Witnesses sworn.]

Mr. SHAYS. Note for the record that the witnesses have responded in the affirmative.

Senator, we're going to start with you. You have as much time as you may want. We're going to turn the clock to 5 minutes, and then do another 5. Bottom line, we want to hear from you. Senator, you're on.

**STATEMENTS OF SENATOR WARREN B. RUDMAN, CHAIRMAN, INDEPENDENT TASK FORCE ON EMERGENCY RESPONDERS, COUNCIL OF FOREIGN RELATIONS; AND AMY SMITHSON, SENIOR FELLOW, CENTER FOR STRATEGIC AND INTERNATIONAL STUDIES**

Senator RUDMAN. Well, Mr. Chairman, thank you; and members of the committee, thank you all very much. I have to tell you I have testified on the Hill now since Hart-Rudman probably 30 times before a variety of committees, and I am going to be very blunt with you. I am really quite taken with your opening statements. I mean, most opening statements are, if you will pardon me, in the Senate and the House opening statements. What I have heard this morning is a keen understanding of this issue which is not surprising when I'd consider the personal history of some of you on the panel who have served in State legislatures, as mayors, as county executives. You obviously have taken a lot of what I was going to say and put it in your statements, and I appreciate that because this is an issue that has nothing to do with partisanship whatsoever. This is an issue that really we have to get our hands around, and we haven't.

Now, as you know, and we testified here over a year ago, Senator Gary Hart and I originally did a study for the Council on Foreign Relations and it ended up saying America's still unprepared, still in danger. It was subsequent to that, the council asked if I would move forward and chair another panel. The reason Gary Hart didn't was at the time he was considering running for President and obviously you can't have anybody in that role who is going to cochair an independent task force. And so I undertook it. If you look at the people on the task force, it's a remarkable assembly of people: two former chairmen of the joint chiefs of staff, head of the FBI and the CIA, the head of the National Football League and the National Basketball Association—why? Because they deal with huge numbers of people in very large, very threatened venues—as well as a Nobel laureate and a number of other very distinguished Americans. The result of that report you have all looked at, obviously, and you are all very familiar with it.

And I would start out and I'm going to summarize very briefly the key recommendations by simply making a statement that I have found no one who could disagree with and that is this: When we sent our young men and women into Iraq—and they are still there—we made sure, to the extent that the technology was available, that they had everything they needed to protect them from any threat that they might face, whether it be chemical, biological or radiological. The Defense Department, with the full support of the Congress, made sure that they had what they needed.

Now, contrast that to the following situation. When something happens in this country, and unfortunately, most people predict it will, it is not anyone in Washington or in the Armed Forces that will respond to that. It is your local police, your local fire, your local EMS people and your hospitals. They will be the front line at least for a while, depending on the extent of the damage. The other thing you have to consider, and this is a rather interesting statistic which was pointed out to me not long ago, and Congresswoman Maloney, I'm sure this will strike home with you, the ratio of people that are killed or injured is normally heavily skewed to injury: Usually 7 to 12 to 1. Not so with the World Trade Center. It was so catastrophic. You had nearly 3,000 people who were killed, and a relatively small number of people who could be taken for treatment.

It is unlikely that kind of a scenario will repeat itself. And so you truly need emergency responders who are able to ameliorate and minimize the kinds of damage that will be caused to maybe thousands of our fellow citizens, depending on the kind of attack, whether it be chemical, whether be it biological, dirty, or conventional nuclear, or even as threatening—large amounts of explosives which are detonated in heavily populated areas can cause enormous damage.

We need only look at what happened in Oklahoma City, to look what happens in Israel, unfortunately, every other week, what happened in Iraq on several occasions, to recognize that it not necessarily only high-technology weapons that cause the threat.

Now, let me point out, as we put out in the report, that this report does not criticize the administration or the Congress. We understand that this is a relatively new issue facing us. This all started in terms of government response on the 11th, although I must say that there were warnings from several groups, including Hart-Rudman and Bremer and Gillmore before the events. Having said that, the government has moved at its usual pace. It's taken a while to get the Department of Homeland Security put together. It is obviously not fully functioning yet, nor should we expect it to. But I'm sure it will.

Congress has the very difficult decision of prioritizing, and so let me start out by saying what I think is the single most important recommendation that we made. If you look at the history of the Department of Defense, the thing they've been very good at is doing threat assessments, and then based on those threat assessments, deciding what capabilities were needed in the various parts of the Armed Forces to meet what those threats were. Then they come up here and they tell the Congress, the House and the Senate, what they think their priorities are. Then you apply your wisdom to it. You negotiate with the administration and you come up with a set of funding priorities.

Now, the problem we have right now is we don't have that. And Dick Clarke and I—who is unfortunately ill this morning and was supposed to be here—testified before the Senate last week and told them what I'm going to tell you here. What we need is a mandate for national minimum standards for homeland security for first responders. Now, we all understand that there are three parts of homeland defense. There is prevention, which is intelligence, and other law enforcement means of preventing. There is protection, which goes to airports and cyber structure, infrastructure, chemical plants, nuclear plants. And there is response. We are dealing only with response in this. But we think it is probably the single most important minimum standard.

We work, by the way, with some remarkable people. You've got two fire chiefs going to testify here, just terrific people we work with from the National Association of Firefighters, the Police Chiefs Association, the National Emergency Responders Association. We work with the National Hospital Association, all the public health officials.

So this report is not a report from 12 or 14 Americans who got together in a room and talked about it. This is what your constituents told us, and we have put together a report which reflects what they think. Now, it is up to you to decide how you want to prioritize.

Congressman Kucinich before he left said you shouldn't spend any of the money on Iraq, you ought to spend it all on homeland security. I'm sure there are many who disagree with that. What you have to decide, because you're elected to decide as I once was, what are the priorities? But you cannot establish priorities until you know what the standards are. What should be the standard in the city of New York for its police, fire and emergency responders?

Now, having said that, let me just go through a few of the points which some of you have already touched on in your opening statements. Some of the findings that we came up with are unequivocal. About half of the fire departments have enough breathing equipment for half of their shifts. That simply means that—or one-third of their shifts. That simply means that two-thirds are totally unable to move into an atmosphere of chemical or biological contamination. But let me point out to you something that we learned in New York. These police and fire people go where they can save lives, even if they put themselves in jeopardy. I think it's terribly unfair to them and their families to send them into jeopardy improperly equipped.

Communication, you all know the story about communication. It's still relatively inoperable. That is not an expensive fix. It ought to be done. That certainly would be one of the standards.

And most of the public health laboratories told us that they do not have the equipment, the expertise, or the personnel to deal with even finding out what we're dealing with. And that is just unacceptable at any level.

And on hazmat, hazardous materials, some of the cities do rather well, but many do not. There have to be standards which, by the way, exist in that area as to what ought to be done.

Now, where are the funds needed? There are huge amounts of funds needed for a national emergency 911 system to be able to get data collection and dispatch, and done accurately and quickly. FEMA has to have money to enhance its urban search and rescue responsibilities, to train local departments to do that, which in a normal kind of an attack would be far more important than it was at the World Trade Center because of the catastrophic collapse of those two buildings.

To foster interoperable communication systems, to enhance the public health laboratories, we believe there ought to be some regional emergency operation centers for local public safety. We believe there ought to be protective gear for WMD remediation equipment to firefighters and police. We think that the national exercises of the type that are going on in Stanford have to be vastly increased so when people have to respond they can respond. We believe that there have to be much better emergency agricultural and veterinary capabilities to deal with contaminated food supplies, and we think there has to be a surge capacity in the Nation's hospitals if there are major attacks.

Now, what were our recommendations, our key recommendations? I'll give you four or five of them.

No. 1, we believe the Congress ought to establish a system for allocating scarce resources based more on addressing the needs rather than on any political basis based on population alone. To do this, you have to consider population density, vulnerability, threat and presence of critical infrastructure.

We believe the House of Representatives ought to transform the House Select Committee on Homeland Security into a standing committee, and we believe the Senate ought to do the same, and it already has a standing committee. But we believe that the consolidation of all of these issues ought to be before the Senate Governmental Affairs Committee.

We believe that Congress ought to require DHS to work with our Federal agency to streamline homeland security grant programs, something that Congressman Tierney talked about. And we heard this over and over again. I mean, you don't necessarily have to fund grants through States or even through counties. There are places where you know what you need, and you can fund it directly to a community with a specific rule for what it will do.

And by the way, the oversight is very important. One of the things that we have seen with block grants in the past is you give a block grant to a community and it goes to supplant their own spending. That's not what this is for. This is to add to what their spending is going to be.

Let me just say, by the way, about the $90 billion figure, we don't know if that's right and we say so. But the reason we don't know is there are no national standards. Once you establish the national standards and then you can determine how much money is being spent at the local, county, and State level, then you can decide what the Federal increment can be. I'm sure it's a great deal of money. It may not be $90; it may be $60, it may be $45. But until you do a set of standards, you just don't know.

We believe that all future appropriations bills that fund emergency response should include strict distribution time lines. There has been just too much delay in getting the money out that's already been appropriated.

Finally, I would say these three are fairly important. We think that you ought to require DHS and Human Services to be part of the standard-setting procedure. They both have important roles to play.

We believe there ought to be an Institute of Best Practices established by DHS to work with State and local governments and other professional associations.

And finally, we think that Congress ought to make emergency responder grants for fiscal year 2004, and thereafter on a multiyear basis, to facilitate long-term planning and training.

I do not envy those of you who sit where you sit. I was on the Senate Appropriations Committee, the Intelligence Committee, a couple of other committees, and I have never seen a situation where prioritization facing appropriators and the Congress in general is as difficult as this one is. I mean, you have major problems in Afghanistan. You have major problems, according to the President, in Iraq. You have a deficit of preparedness that is serious.

And let me just close by saying this. You know, I've been in politics for some time, both at the State and the national level. I think I understand something about the response of individual Americans to individual actions. I come from a small State which is roughly a million or 1.1 million people. You tend to get a feel for people pretty well in a State like that. Probably you do in the House. Probably less so in the Senate where you come from a large State.

Mr. SHAYS. There's no doubt about that, Senator.

Senator RUDMAN. Well, you only represent half a million people, right?

Mr. SHAYS. I just want to say, there's no doubt about that.

Senator RUDMAN. Well I said it, so you're agreeing with me.

I'll make this observation: that if there's another attack—and I think there will be, it's only a question of time—and we have a catastrophe, we are unable to help our citizens after all of the warnings we have received, I will put it bluntly, there will be hell to pay for those people who are policymakers, including all Members of Congress, whether you did the right thing or you didn't do the right thing. And so from a point of view of doing the right thing and what's politically right, this is high priority, and I get the sense talking to Members of Congress on both sides of the Capitol that people recognize it. So you're going to have to decide.

Thank you.

Mr. SHAYS. Thank you for that very thorough and helpful statement and thank you for all your good work.

[NOTE.—The report entitled, "Emergency Responders: Drastically Underfunded, Dangerously Unprepared," may be found in subcommittee files.]

Mr. SHAYS. Dr. Smithson, you're recognized. If you compliment this committee, the clock becomes irrelevant.

Dr. SMITHSON. Delighted to hear that.

Good morning, Mr. Chairman, members of the committee. Like you and certainly like the council's task force, I've spent a great deal of time learning from America's first responders of all disciplines. Perhaps that's why I'm in such agreement with the council's report on emergency responders.

After domestic preparedness programs were initiated by Senators Nunn, Lugar, and Domenici in 1996 to start addressing this problem, I began research on how these programs were working. In fact, I interviewed responders from 33 cities and 25 States. My research was presented in October 2002 in a chunky report called "Ataxia."

In the last few years, far too many from the front lines continue to tell me that they have yet to see a dime of the money that the Federal Government is spending on terrorism preparedness. Since the key to domestic preparedness—and I think I'm hearing agreement on this—lies in improved local response capabilities, I applaud the council's recommendation to increase dramatically Federal spending to locals. In fact, to find a major point disagreement with the council's task force, I have to do something slightly unfair, and my apologies, Senator Rudman. I have to resort to their 2002 report which recommended tripling the number of National Guard Civil Support Teams. Given the astuteness of their other recommendations, the council's backing of this politically popular placebo program is somewhat disappointing.

Here's what the frontline responders, including those serving in the National Guard, had to say about these teams. And I convey their views with utmost respect to our men and women in uniform. Frontline emergency professionals have an unmistakable message about these teams. They have minuscule, if not negative, utility in disaster response, and the resources that they consume could be much better applied locally. Though stocked with top-of-the-line equipment, Civil Support Teams have practically zero actual emergency response experience, which explains why frontliners can regale me with tales of how these teams have botched it during exercises.

One deputy emergency manager bluntly told me, "The good thing about these teams is that it takes them as long as it does to get here."

On that point, Congresswoman Maloney, the New York Civil Support Team arrived at the scene roughly a dozen hours after planes struck the World Trade Towers and initiated monitoring that was already being done by New York City and U.S. environmental authorities.

The dynamics of chemical disaster response are such that the Guard's teams cannot arrive in time to save lives. Moreover, their four-person medical component will barely register in a biological disaster.

Consider that the combined $9.4 million that it costs to stand up and equip a Guard team with chemical and biological gear could instead be used to purchase 6,600 emergency power generators, or to buy level A personal protective gear for 3,700 firefighters so that they could safely enter hot zones. With the $176 million required to maintain all 55 Civil Support Teams each year, over 586,600 police officers could be equipped with high-performance masks that would enable them to stay on duty if terrorists use unconventional weapons.

My point, as the chart in my written testimony amply shows, is that were these moneys instead invested in equipment for frontline responders who would be at the scene of a disaster in minutes rather than in hours, America would be better prepared to grapple with terrorist attacks.

Washington can continue to go about this in an expensive and inefficient way, or Congress can get America on the smart route to enhance terrorism and disaster preparedness. I think we're all in agreement about the need to get money to the front lines. But Congress should also direct that exercises be conducted on a no-notice basis and that the after-action reports from these drills be exempt from Freedom of Information Act requests. Doing so would allow response deficiencies to be identified and corrected, and that's not what's happening with the process as it currently exists.

Furthermore, I would encourage Congress to mandate that the executive branch get swiftly and diligently to work with local responders, professional organizations, and Governors to develop and institutionalize preparedness standards, just as the council recommended and as I did a few years ago. You should also insist that burden-sharing arrangements be crafted between Federal, State and local authorities so that preparedness can be sustained.

You're not the only ones that have heard already we're beginning to backslide on the advances that we've made in preparedness. So in chapter 7 of "Ataxia," I made several recommendations for cost-sharing alternatives. By formally consolidating congressional oversight and exercising these responsibilities vigorously, you can eliminate redundant and poorly designed programs.

I have a number of recommendations and information from the front lines about how we can get more effective communications, and I would also let you know that the council's recommendation to have first responders get ready access to best practices is actually being put into play by the Memorial Institute for the Prevention of Terrorism. Until recently, I participated in a project that's about to launch to share best practices and lessons learned.

In closing, if you haven't already, please drop by your local fire station, police station, emergency operations center, hospital or public health lab and ask frontline responders how to streamline Federal programs and get this country better prepared. They'll give you an earful. They'll tell you what programs are clunkers and what's working right. I urge you to listen to them closely to heed their counsel before you cast votes on homeland security issues in the days, months, and years ahead.

Thank you.

Mr. SHAYS. Thank you very much.

[The prepared statement of Dr. Smithson follows:]

Many date the nation's terrorism preparedness efforts to the aftermath of September 11, 2001. Indeed, the last two years have witnessed the launching of a number of high-profile efforts, including the establishment of the homeland security advisory system, major increases in federal spending on terrorism prevention and preparedness, and the creation of the Department of Homeland Security. However, significant efforts to ready America's front-line emergency personnel to contend with terrorist use of unconventional weapons really began when Senators Richard Lugar (R-Indiana) and Sam Nunn (D-Georgia, ret.) seized on Aum Shinrikyo's 20 March 1995 sarin gas attack in Tokyo as a harbinger of possible terrorist attacks in the United States. With the Domestic Preparedness Programs segment of the 1996 Defense Authorization Act, they and Senator Pete Domenici (R-New Mexico) kicked off a series of training and equipment programs to increase the ability of America's largest metropolitan areas to confront unconventional terrorist attacks.

In its report "Emergency Responders: Drastically Underfunded, Dangerously Unprepared," the Council on Foreign Relations terrorism task force rightly points out that a great deal of work remains if front-line US responders are to be as ready as possible to contend with the disasters that terrorists or, more likely, Mother Nature could bring their way. Not only is the Council's report accurately titled, the major recommendations of this report are right on target. I say this with conviction because when I delved into the state of preparedness on America's front lines with research involving responders of all disciplines from 33 US cities in 25 states, I found them drastically underfunded as well. The shortfalls in emergency response funding can be seen in US cities, but budget tallies tell the tale as well. According to the Office of Management and Budget, of the total $8.4 and $8.7 billion that the US government spent on defense against terrorism in 2000 and 2001, respectively, only $315 and $311 million reached front-line US responders in the form of training, planning, and equipment grants in those two years. Since the publication of *Ataxia: The Chemical and Biological Terrorism Threat and the US Response* in October 2000, I have continued to interact regularly with the full spectrum of emergency personnel from a great many US cities.

Congress and the Executive Branch should take note that many firefighters, police officers, paramedics, public health officials, health care providers, and emergency managers have told me they have yet to see a dime of the federal terrorism prevention and preparedness monies appropriated since September 11$^{th}$. Not only have federal terrorism preparedness dollars not trickled down to many front-line responders, local budgets that are so tight that fire chiefs cannot replace worn out respirators and public hospital administrators are pinching pennies to buy new gurneys. While insufficient funding partially explains why America's front-line responders are not as prepared as they could be, thank goodness they will do their utmost to save the lives of their fellow citizens should calamity strike, even if they lack the best equipment and are not trained in the latest response techniques.

Public safety personnel and health care providers risk their lives during disasters like the September 11$^{th}$ attacks, the 19 April 1995 bombing of the Murrah Federal Building in Oklahoma City, and the 17 January 1994 Northridge earthquake, just as they do on a daily basis when armed robberies, hazardous cargo accidents, and tornados harm US communities. For that reason, those who set policies and make decisions about how to spend US tax dollars owe emergency response personnel and all Americans more purposeful and well-conceived terrorism prevention and preparedness programs that focus on the front lines instead of inside the beltway. At the very least, as the Council's task force observes, Congress should dispense with evenly dividing the "spoils," not to mention its pet programs, and make terrorism preparedness funding decisions instead based on population density and the comparative vulnerability of various locations to terrorist attack.

So, to begin with, the key to domestic preparedness lies in getting taxpayers' dollars channeled to readiness at the local level, where improved response capacities will better arm public safety and medical personnel to contend with disasters, whether natural, accidental, or intentional. Therefore, I applaud the Council's recommendation to increase dramatically federal terrorism prevention and preparedness spending at the local level.

**Charting A Practical Course for Long-term Terrorism Readiness**
Well-considered investments are better than hasty ones, and US policy makers made numerous major spending decisions rapidly after September 11th. The Council on Foreign Relations found, as did I in research for *Ataxia*, that certain approaches are needed to get the most preparedness bang for US tax dollars. Newcomers tend to think that if only emergency personnel had the right equipment, they could respond well. True, equipment is an important part of the response equation and more funds are needed for personal protective gear, field detectors, laboratory analysis equipment, and antidotes, among other items, but emergency personnel also need to be well trained, including the regular practice of skills and use of equipment in field exercises.

Should Congress inquire it is likely to find, like I did, that a relatively small fraction of US front-line response personnel have made it into the classroom for any dedicated or in-depth terrorism response training. In particular, far too many health care givers, 9-1-1 call dispatchers, and police have not received terrorism response training. Moreover, as the Council's terrorism task force also recognized, little of the training provided to date has involved guidelines or standards that the different response disciplines agree should be institutionalized and upheld by practitioners.

Institutionalization of professional standards is the time-tested and commonsense approach that underpins the all-hazards, echelons-of-response system that both states and cities know and advocate. If preparedness is truly to take hold nationwide on the front lines and be sustained in perpetuity, then appropriate terrorism response standards belong in the local and state training academies, as well as in the nursing and medical schools. Institutionalization is the most cost-effective way to spread training geographically and build a tiered response capability.

Roughly seven years into the domestic preparedness effort, the time has come for Washington to turn training over to the appropriate professional and local entities that will take preparedness forward more systematically and cost effectively. No overarching structure is in place, however, to move any of the entities concerned smartly forward to create and perpetuate terrorism response standards. Given the advantages that institutionalization offers, I stated in Chapter 7 of *Ataxia* that Washington could best demonstrate its seriousness about nationwide preparedness by working diligently with local responders, professional organizations, and governors nationwide to develop, roll out, and institutionalize standards according to an agreed time line. The federal government's role is to be the catalyst and convener that prods the tangle of entities to get this important job done. Until that occurs, training lacking in standards will be implemented unevenly, in pockets. Perhaps even a decade hence, the majority of US responders will still not be appropriately trained. Specification of standards and institutionalization of training clearly make more sense than that.

*New Governing Rules for Exercises*

Regular exercises, both in the field and so-called tabletop decision-making drills, are essential to preparedness. If specialized equipment remains parked in a warehouse and emergency personnel do not test seldom-used plans and skills, then preparedness atrophies. Exercises can be large, extremely orchestrated, pre-notified affairs such as TOPOFF I and II, in which case their utility often lies less in their actual conduct than in the extensive preparations prior to the event. Leading into a big exercise, local personnel meet frequently to revise plans and capabilities so that they look their best during the exercise. To a lesser extent, this type of polishing also occurs with smaller local drills.

Exercises would be more valuable learning experiences if two important adjustments were made in how they are conducted. First, the planning and polishing that takes place before an exercise is needed and productive, but to obtain a more realistic test of response preparedness, the exercise itself should be initiated with no or as little notice as possible. A no-notice exercise means that in larger drills, federal assets would not be pre-picked and pre-staged, like they have been in TOPOFF and other drills involving federal personnel. The terms of the exercise should specify that teams deploy as notified. While the general nature and identity of the exercise location(s) would certainly be known beforehand and the timeframe of the drill agreed within a window of several months, a handful of local officials should have the discretion to trigger the onset of exercises. This approach would require everyone to dispense with the comfortable claims of what they could do and reveal more about what they actually can do. A more genuine and probably sobering measure of federal capabilities could be taken, and the outcome of the exercise could better inform federal, state, and local plans, programs, and capabilities.

More often than not, however, the real lessons of exercises large and modest are not accurately conveyed, a problem that points to the need for a second reform of the rules that govern exercises. Because exercise after action reports are subject to Freedom of Information Act requests, they are written so that they smooth over the problems that the drill uncovered, showing local, state, and federal response agencies in their best light. Emergency response officials have been burned unfairly when politicians and press access and use exercise reports for political gain and to generate sensation headlines. Local response personnel note that the other reason that after action reports often are not worth reading is that contractor-prepared reports tend to be vapid, cookie-cutter documents. These circumstances stymie the learning process and therefore undercut opportunities to improve response plans, practices, and capabilities after drills.

To obtain the full preparedness benefit from exercises conducted with federal dollars, Congress should consider making after action reports exempt from Freedom of Information Act requests. In addition, Congress should weigh a requirement that experienced, non-local, professional responders serve as exercise observers and lead the "hot wash" evaluations immediately after drills that form the foundation of exercise reports. Absent such reforms, front-line responders will continue to experience lessons the hard way, time and again, but they are unlikely to learn from their experiences and make the appropriate institutional changes to improve response plans, capabilities and practices.

*Capturing Lessons Learned and Best Practices*

In a related point, the Council on Foreign Relations urges Congress to establish within the Homeland Security Department a National Institute for Best Practices in Emergency Preparedness that gives emergency responders access to best practices and lessons learned via the worldwide web. One of the things that emergency personnel have told me most frequently is that they wanted to benefit from the experiences and innovations of their fellow responders. As Chapter 6 of *Ataxia* documents, emergency response personnel everywhere are confronting similar obstacles. In some jurisdictions, emergency response officials have made noteworthy progress in scaling some of the significant hurdles associated with chemical and biological disaster response. Others who are still struggling to put various response systems and capabilities in place could save time and resources if they understood how their contemporaries figured out the way to: harden possible targets against terrorist attack; overcome the decontamination bottleneck at hospitals; provide mass prophylaxis with a minimum of health care workers; get hospitals to work together regionally; take a cheap shortcut to comprehensive, redundant emergency communications; or institute syndromic surveillance so that leading edge indicators of a disease outbreak can alert emergency personnel to a covert bioterrorist attack or a naturally occurring eruption of disease. Sharing this type of knowledge will enable emergency response personnel across the country to get better prepared more rapidly than if they had to recreate the wheels already discovered by others.

Until recently, I was involved in just such a best practices/lessons learned project, sponsored by the Memorial Institute for the Prevention of Terrorism in Oklahoma City. If current plans hold, the Memorial Institute will soon debut a website that captures the type of knowledge that responders seek, namely the details of why, how, and what emergency personnel did to develop these practices; the unintended consequences of establishing these practices; how problems encountered were handled; and, the modifications made to the nascent practices to improve them. The Memorial Institute's project could well fulfill the objectives of the Council's recommendation by collecting and propagating via a secure website at no cost to front-line personnel the cutting edge, tricks-of-the-trade knowledge that countless responders have told me they want and need.

*Hammering Out Burden-sharing Arrangements*

Yet another major point that the Council's report raises is the need for federal, state, and local governments to come to terms about responsibilities and commit the fiscal resources to meet them. The original Domestic Preparedness Programs were structured as a cost-sharing arrangement, such that the federal government provided training and equipment while city governments paid local labor costs. Even after September 11th and given the press of requirements to provide daily governmental services, it can be an uphill battle to get response agency chiefs, city councils, hospital officials, and mayors to authorize the overtime labor costs for training and the other expenditures that accompany terrorism preparedness (e.g., equipment maintenance). One city emergency manager told me that preparation for an unconventional terrorist attack rated "somewhere below the likelihood of a tsunami and a step above an alien invasion."

An important part of sustaining preparedness concerns how to pay the bills for capital improvements, equipment replacement and maintenance, and exercises. The Council on Foreign Relations is right to stress the urgency of having federal, state, and local authorities settle this matter. Front-line emergency personnel tell me that preparedness gains have already begun to degrade in some cities. Unless a long-term cost-sharing arrangement is created, both the local and federal investments to date could slide into reverse. In Chapter 7 of *Ataxia*, I proposed such cost-sharing alternatives as ongoing federal funding, state and local disaster preparedness trust funds, and local user fees. A popular saying among politicians is that no time is the right time to discuss higher federal budgets or new local taxes. Hopefully, September 11th changed the willingness of federal, state, and local officials to begin this long overdue discussion.

**Inside Vs. Outside the Beltway Investment Perspectives**

To find a major point of disagreement with the Council on Foreign Relations task force on terrorism, I have to resort to last year's task force report. In "America---Still Unprepared, Still in Danger," released in October 2002, the Council recommended that Congress triple the number of National Guard Civil Support Teams. Given the astuteness of their other recommendations, the Council's backing of this politically popular placebo program was disappointing. I again urge Congress to consider the evaluation of these teams offered by the front-line responders, including some serving in the National Guard, whom I interviewed for *Ataxia*. I convey their views with utmost respect for the service that uniformed men and women, both active duty and reserve, perform for our country. Furthermore, I would note that local emergency response personnel stated their intent to call on the National Guard for the types of support that the Guard has traditionally provided so well after disasters, such as help with logistic and public safety missions.

However, with regard to the National Guard Civil Support Teams, the message from the front line is unified and clear: They have minuscule, if not negative, utility in disaster response, and the resources that they consume could be much better applied locally, where they could make a real preparedness difference. The Civil Support Teams are stocked with top-of-the-line equipment and are often trained by front-line responders, but they have practically zero emergency response experience. Lack of *bona fide* emergency response experience is why local veteran responders are reluctant to substitute the advice of Civil Support Teams for their own, seasoned judgments. In the mid-May 2000 TOPOFF exercise, the Civil Support Team in Denver insisted that it had identified the mystery biological agent with SMART tickets, which have such high false positive and false negative rates that numerous cities refused to buy them. The team in Portsmouth lacked the technical expertise to understand the minimal hazard posed by mustard on a chilly, 49-degree day. To old hands at epidemiological investigations and hazardous material operations, the absurdity of these two anecdotes is readily apparent. The deputy director of one city's Office of Emergency Management bluntly told me that "The good thing about those teams is that it takes them as long as it does to get here."

On that point, the New York Civil Support Team arrived at the scene roughly a dozen hours after planes struck the World Trade Towers and proceeded with environmental monitoring that was redundant of efforts undertaken hours earlier by New York City agencies as well as the US Environmental Protection Agency. The dynamics of a chemical disaster response are such that the Guard's teams cannot arrive in time to make a life-saving difference. In the moments after Aum Shinrikyo's sarin gas attack against the commuters in Tokyo's subway system on 20 March 1995, local transit workers, police, firefighters, and health care providers aided people gasping for air, some in need of quick administration of the nerve agent antidotes to save their lives. The attack unfolded from 7:46 to 8:01 am. The first patients reached the nearest hospital less than 30 minutes later. The Japanese Self Defense Forces dispatched its special chemical defense units downtown at 10:10am. Although these units were located in Tokyo's outskirts, the teams, caught in huge traffic jams, did not reach the attack scene until two and a half to roughly five hours later. The victims of the attack had long since been cleared from the scene. Not only will Civil Support Teams be challenged to reach the scene of a disaster any sooner than the Japanese Self Defense Forces did in Tokyo, their applicability in a biological disaster is truly marginal. The four-person medical component of the Guard's Civil Support Teams is a drop in the bucket of the medical personnel that would be needed in an experiencing a major disease outbreak.

To those accustomed to overseeing billion dollar budgets, this National Guard Civil Support Team program might not seem so ill advised. Please consider, however, how this program's budget could be put to uses that would make a measurable preparedness difference on the front lines. As the chart below indicates, the National Guard states that it costs over $2 million to equip a Civil Support Team with chemical and biological detection gear, $7.4 million to stand up---organize and train---a team, and $3.2 million per year to maintain a Civil Support Team. A total of $176 million will be needed each year to maintain the full compliment of fifty-five Civil Support Teams. Were these monies instead invested in various equipment items for front-line career and volunteer responders who could be at the scene of a disaster within minutes rather than hours, then arguably America would be better prepared to grapple with terrorist attacks. For example, police officers are greatly concerned because they do not have appropriate respiratory protection to allow them to stay on the beat if terrorists employ unconventional weapons. With $176 million, over 586,600 police officers could be equipped with a high performance mask that has canisters to filter extremely toxic industrial and even warfare chemicals, such as the MSA 1000 CBA-RCA mask. Other examples of how these funds could be spent are listed below, working from the high end of the cost estimate ranges for items found on Standardized Equipment List published by the Office of Domestic Preparedness.

**Comparative Investments: National Guard Civil Support Teams vs. Equipment for Front-line Responders.**

| | National Guard Civil Support Team Costs* | | | |
|---|---|---|---|---|
| | **Equipment Costs Per Team** | **Stand-Up Costs Per Team** | **Yearly Maintenance Per Team** | **Total Yearly Maintenance for 55 Civil Support Teams** |
| | $2,000,000 | $7,400,000 | $3,200,000 | $176,000,000 |
| **Minimum Number of Selected Items That Can be Purchased for the above amounts**** | | | | |
| Front-Line Disaster Response Equipment** | | | | |
| Level A Protective Equipment (certified, reusable, disposable) | 1,000 | 3,700 | 1,600 | 88,000 |
| Powered Air-Purifying Respirators (tight-fitting, full face piece, with chemically resistant hood, appropriate cartridges) | 3,333 | 12,333 | 5,333 | 293,333 |
| Boundary Marking Tape (yellow, red, etc.) | 100,000 | 370,000 | 160,000 | 8,800,000 |
| Air Compressors (to refill respirator canisters) | 6,667 | 24,667 | 10,667 | 586,667 |
| Generators | 6,667 | 2,467 | 1,067 | 58,667 |
| Mask Leak/Fit Testers | 67 | 247 | 107 | 5,867 |
| **Minimum Number of Selected Items That Can be Purchased for the above amounts**** | | | | |
| Helmet-Mounted Lighting Systems | 133,333 | 493,333 | 213,333 | 11,733,333 |
| Tactical Body Armor | 667 | 2,467 | 1,067 | 58,667 |
| Cooling Garments (to manage heat stress) | 4,000 | 14,800 | 6,400 | 352,000 |
| Fire Resistant Gloves | 33,333 | 123,333 | 53,333 | 2,933,333 |
| Robots (basic bomb mitigation, remediation) | 13 | 49 | 21 | 1,173 |
| Lifting Devices (air bag systems, hydraulic rams, jacks, ropes, etc.) | 133 | 493 | 213 | 11,733 |



92-127 D- 3

| | | | | |
|---|---|---|---|---|
| Bull Horns | 18,182 | 67,273 | 29,091 | 1,600,000 |
| Land Mobile, Two-way In-Suit Communications (secure, hands-free, etc.) | 400 | 1,480 | 640 | 35,200 |
| Multi-Channel Radios (encrypted) | 400 | 1,480 | 640 | 35,200 |
| Gas Chromatograph/ Mass Spectrometer | 100 | 370 | 160 | 8,800 |
| Portable Biological Air Sampler | 167 | 617 | 267 | 14,667 |
| Liquid, Solid, or Air/Vapor Chemical Sampling/ Evidence Kits | 2,000 | 7,400 | 3,200 | 176,000 |
| Pressurized Sprayers (for decontamination) | 4,000 | 14,800 | 6,400 | 352,000 |
| Traffic Cones, Directional Signage (in multiple languages/ pictographs) | 4,000 | 14,800 | 6,400 | 352,000 |
| Video Assessment/Cameras | 4 | 15 | 6 | 352 |
| Body Bags (heavy-duty) | 20,000 | 74,000 | 32,000 | 1,760,000 |
| Disposable Emergency Blankets | 666,667 | 2,466,667 | 1,066,667 | 58,666,667 |
| Sterile Dressing (assorted sizes) | 100,000 | 370,000 | 160,000 | 8,800,000 |
| Endotracheal Tubes | 66,667 | 246,667 | 106,667 | 5,866,667 |
| Cyanide Antidote Kits | 6,154 | 22,769 | 9,846 | 541,538 |
| Doxycycline (TAB 100 mg 500s) | 25,000 | 92,500 | 40,000 | 2,200,000 |

Notes: * Lt.Col. Thomas Hook, chief of the National Guard's Civil Support Team program provided costs for various aspects of the program to Morgan Courtney of CSIS in a 5 September 2003 conversation.
** Prices of equipment items taken from the Office of Domestic Preparedness' State Domestic Preparedness Equipment Program Standardized Equipment List. A price range (e.g., generators for $400-$3,000 each) was frequently given. The number items that could be bought was calculated based on the highest cost estimate for each particular item.

The refrain heard inside the beltway when the National Guard or federal response teams are criticized as redundant and unable to reach the site to accomplish their asserted missions is that creating and enhancing teams does not really cost much—just a few million dollars here and there. A million dollars may be pocket change in the Pentagon's budget, but it is serious money on the front lines. Moreover, a few million poorly spent in several programs adds up to a tidy lump sum. As the table above shows, hundreds of thousands of front-line responders could be better equipped if Washington were to ax just the National Guard Civil Support Team program. Imagine what could be done with the funds if Congress halted other redundant, ill-conceived terrorism response programs.

**Concluding Observations**

The pragmatic steps to better local terrorism preparedness are clear. Washington can take the smart route to enhance terrorism and disaster preparedness nationwide or it can continue to go about this in an expensive and inefficient way. National preparedness lie not in more federal bureaucracy, but in such commonsense policies and programs such as:

- The bulk of federal funds need to be devoted to multiyear grants that enhance readiness at the local level, where an increase in skills, training, and equipment would make a genuine life-saving difference. Even if terrorists never strike again in this country, such investments would be well worthwhile because they would improve the ability of hometown rescuers to respond to everyday emergencies.
- Drills at the local, state, and federal levels are necessary because plans, equipment, and skills that are unused for extended periods of time often do not work fully when emergencies occur. Exercises would be more worthwhile if they were conducted on a no-notice basis and candidly evaluated without fear of undue penalties for poor performance, thereby allowing response deficiencies to be corrected.
- Appropriate steps should be taken to see that all frontline response disciplines benefit from the development of professional standards and the institutionalization of terrorism response standards in the nation's training academies, universities, and schools.
- Federal, state, and local officials need to develop and move forward with a plan to share costs and thereby sustain preparedness over the long term.
- Congress should exercise its oversight responsibilities vigorously, eliminating redundant and poorly designed programs and diverting those funds instead to preparedness efforts at the local level.
- Last, but certainly not least, an essential element of streamlining and coordinating government programs lies here, in Congress. So long as congressional oversight is fractured, individual federal agencies may continue to exploit the situation to the advantage of their own institutional interests and the detriment of coordinated, cost-effective programming. Congressional oversight sorely needs to be consolidated formally, as the Council also suggests, in the House Select Committee on Homeland Security and the Senate Government Affairs Committee.

Those who know first-hand the tremendous demands of responding to a disaster have a saying: "All emergencies are local." Heroic actions following the 1989 Loma Prieta earthquake, Aum Shinrikyo's 1995 poison gas attack, and September 11$^{th}$ underscore the basic truth of who saves lives when natural or manmade calamity strikes. In such circumstances, the lifesavers are not federal response teams that swoop in from across the country, but the local firefighters, police, paramedics, nurses, and physicians.

The soundest policies are based not on book learning and hypotheses, but rather on real life experience. As is well known, real life is what exists outside of Washington's beltway. When it comes to crafting the best policies and programs to enhance this nation's ability to prevent and respond to terrorist attacks, Congress should seek out, learn from, and be guided by this nation's most experienced emergency planners and responders. The next time that you are in your home districts or, for that matter, anywhere else in the country, drop by a fire or police station, a hospital, a public health laboratory, the public works department, or an emergency operations center and ask front-line emergency personnel to explain the commonsense keys to streamline federal programs and cut costs while improving local abilities to prevent, mitigate, prepare for, and respond to terrorist attacks. Mr. Chairman, Members of the Committee, I close asking that you listen closely and heed the counsel of America's best, brightest, and bravest emergency response professionals before you cast votes related to homeland security in the weeks, months, and years to come.

Mr. SHAYS. At this time we'll begin with our questions, and given the number of Members, I think we'll just do 5 minutes and we can go on to a second round. It's going stay up there. Mr. Turner, you have the floor.

Mr. TURNER. Thank you. I'm asking a question about the standardization process. As I said in my opening statement, serving as a mayor of a city, a city that had a weapons-of-mass-destruction exercise prior to September 11, the issue of standardization is an issue that I think is going to be a very difficult process.

Let me tell you the story of Dayton. We had a weapons-of-mass-destruction exercise at an arena. We even closed part of an interstate highway, went through a mock decontamination process. But the most important thing that we learned during all of that was the coordination between the different organizations throughout the region—the FBI, we have Wright Patterson Air Force Base, the military, the county and the State officials—so that when September 11 happened, we were a community that went and dusted off our books as to what our plan was, and deployed it, and it was a marvelous thing to see because we didn't run around saying, who's in charge, what should we do? People knew what streets to close. They knew who they needed to coordinate with. That's one aspect of preparedness.

When we talk about standardization of preparedness and what first responders need, Senator, I found your comments interesting about the difference between the population and the actual needs of the community. To give you one example, in my district there's a small town, Wilmington, OH, that has Airborne Express in its back yard. Over a million packages go through a sorting facility there every day. And the—in Wilmington, OH, their hospital does not have sufficient emergency response equipment to respond if something should happen at that facility. So they're trying to come up to speed. If you did a strictly population-based analysis, you would get to Wilmington, OH sometime in 2020.

And then if you look at the various other communities that are around, they all have differing needs. You know, St. Louis, with the arch, has different needs than Wright Patterson Air Force Base in Dayton, OH. How do we go through a standardization process that actually advances our preparedness, that we don't get caught in a quagmire of trying to overanalyze, but at the same time that we don't miss a bunch?

Senator RUDMAN. Yeah. Our answer would be generally as follows. First, a threat assessment done properly from the Federal Government, on down through the various States, counties, cities, would identify high-threat targets. And you've mentioned one. That's a place where something bad could happen in one of these packages that's being delivered. So you have to first decide where the threats are.

Now, obviously, Congressman Maloney is right. New York City, Washington, it's very obvious for a lot of reasons. There are other major cities as well. However, there are smaller places that have critical infrastructure of various kinds that have to be identified. So you have to put those in the matrix of where the threat assessment comes out. Once you do that, the kind of equipment required for various places varies on what the threat and the population and the mobility is in a particular area. For instance, the State that I come from, New Hampshire, could probably have a series of regional units that could combine under certain circumstances, whereas a large city like Dayton or Syracuse, NY, or Bridgeport, CT would probably need much of that equipment. The standardization of the equipment for these people is not hard to come up with. We already have been told by them what they think a minimum standard is, and that will have to be studied in depth to see if they're high or low.

The point you're raising is how does the threat assessment combine with that equipment need. And my answer would be, thinking about it here this morning, that would depend entirely on what the minimum standard would be, and then do certain communities need more rather than less? But in terms of the police, fire, and emergency responders, the basic kind of equipment they need to be able to do their job is a standard that should not be hard to set. The hardest thing will be to decide what you need in a particular community, and that will take a lot of work by a lot of people. But we think it could be done in a 6-month to 1-year timeframe.

Dr. SMITHSON. Congressman, you're right. The effort to create standards is going to be rather complicated. But, there are several professional organizations that have a running start on this. And if you just ask the locals and do some exploration as I did, you'll find a number of professional practices that are emerging that can grapple with some of the tough response issues. If we can get agreement on these professional standards, the way to move this thing forward is to put these standards, to have the Governors agree to put these standards in local police and fire academies, and to get medical schools and nursing schools to change their curricula so that future generations are taught, and we're not always behind the wheel here. Otherwise, a decade from now, we'll still have defense contractors out there teaching various practices here and there. We need to do this in a way that is cost effective and that gets this information promulgated across the country. These military teams that won't get there first. I think you already recognize that. It will be the neighbors next door that get there first to help you. And these are the very people that need the standards in their academies and in their colleges and universities.

Thank you.

Mr. SHAYS. I thank the gentleman. I thank the witnesses.

Mrs. Maloney.

Mrs. MALONEY. Thank you, Mr. Chairman. What about your time?

Mr. RUPPERSBERGER. He'll get time. He's the chairman.

Mrs. MALONEY. OK. Senator Rudman, the only comment I have heard from the Department of Homeland Security about your very excellent report is that the money that you call for would be used to buy gold-plated telephones. And to your knowledge, have they gotten in touch with you or with anyone on the council to discuss your very thoughtful report in depth?

Senator RUDMAN. Congresswoman Maloney, let me just respond to that comment. You know, unfortunately—and I don't mean to be hard on the public relations profession—but every government agency and, quite frankly, every congressional office has a spokesman, and they're capable of putting out the most incredible dribble. That's the only word that I can use. I mean, their natural response is not to even look at what somebody said but to immediately go into a full defensive mode. Secretary Ridge was pretty hard on that statement on "Meet the Press," and I'll take that as being the final word. That was a stupid statement. We weren't looking for gold-plated telephones. We'd like the New York City fire and police departments to be able to talk to each other.

Mrs. MALONEY. Has he reached out, regardless of the inappropriate statement, has he reached out to speak with you in the Department about the report?

Senator RUDMAN. Some of our staff have talked to some of their staff. They have not contacted me personally. Nor did they after the last report we did.

Mr. SHAYS. Will the gentlelady yield? You know that will be a good basis for us inviting some people from the Department of Homeland Security in to just talk about this report, and maybe if someone from your staff wants to participate, we would allow that as well.

Senator RUDMAN. We would be happy to do it. I mean the fact is that I have met at great length with John Gordon, who's now got Tom Ridge's job in the White House, General John Gordon, retired four-star. We've talked about the report a great deal from the White House point of view. I have not personally talked, but people from our group have talked to staff at DHS.

Mrs. MALONEY. I would like to go back to one of your comments on fairness and going where the threat is. And press reports this year highlighted the unfairness, really, of the homeland security grant distribution program, and one report showed that under the latest version passed by the Senate, Wyoming would be getting $32.25 per person, while New York City, by all accounts, by all analysis, target No. 1, would only get $4.60 per person. And we've heard talk from the administration about changing the formula. As Secretary Ridge has publicly said, he does not support this. He thinks it should go where the threat is. But we've seen nothing really of substance besides rhetoric.

And what is your opinion of this grant distribution formula? And what is your advice on how we can get it changed in a way that really focuses on where the threat is, because everybody says, "Oh we should." But when you look at the paperwork that's moving through Congress, they're not. They're going on a formula that really is more of a political formula than a threat formula.

Senator RUDMAN. We totally agree with you. I believe Secretary Ridge agrees with you.

Mrs. MALONEY. He does publicly.

Senator RUDMAN. And we have laid out in our report the metrics of how we believe you ought to proceed. It is based, first, on a threat assessment and then based on the standards that you set, and then apply those to the various communities to see what they need. Obviously, you need a higher density of expenditure for New York City then you do for Manchester, NH. It's a whole different situation. And it cannot be done the way it's being done presently, because then you will be essentially putting more money where it's not needed and less money where it is needed. So——

Mrs. MALONEY. So you would start off with a threat assessment.

Senator RUDMAN. The threat assessment, the standards, and a distribution formula based on that.

Mrs. MALONEY. Threat assessment, the standards, and distribution base on that.

Senator RUDMAN. And I might tell you that this government is very good at doing that. I mean, there are plenty of systems analysts who are employed in this government who know exactly how to do that. But this appears to me to be a very hastily thrown together formula for distributing money.

Mrs. MALONEY. It's a political formula.

Senator RUDMAN. I believe it is.

Mrs. MALONEY. It's not—and we were speaking yesterday with Doug Duncan, who is the Montgomery County, MD executive, and he told us that in order to prepare his county for terrorism, he has had to cut the budget of education and other social services. And have you heard similar stories around the country and does it surprise you?

Also Dr. Smithson, I would love a copy of your report. I haven't read that one, although I've read this one several times.

Senator RUDMAN. We have heard that and I would also point out something that has not been mentioned here this morning. I don't know if you're aware of this or not, but we now have less policemen in this country than we did on September 11. Now, that's hard to believe. There are less police due to budget cuts across the country than there were on September 11.

Mrs. MALONEY. Based on whose stats?

Senator RUDMAN. We have the statistics from the National Association of Chiefs of Police and other places.

Mrs. MALONEY. I'm out of time.

Senator RUDMAN. And I believe New York is one of those places.

Mrs. MALONEY. Well, police and fire are less than what they were.

Mr. SHAYS. Thank you. Mr. Murphy.

Mr. MURPHY. Thank you, Mr. Chairman.

Senator, I am particularly interested in this report, some of the issues involving hospital and health care providers, because I think that when we talk about first responders we usually think paramedics, police, and firemen and getting to the scene. But, of course, the next step is getting them to the hospital in a coordinated way. And I know in my region around Pittsburgh, some of the concern is how we communicate with tens of thousands of nurses and physicians, tell them where to go and how to get there and what to be ready for when they do get there. Now—and thinking how to use paging systems, BlackBerry systems, personal devices, phones, etc. Now, when we look at this other level of what happens if there was a power grid meltdown, how do you get ahold of anybody? Any suggestions of what should be done in that sort of a preparedness?

Senator RUDMAN. One of the things that you will find in the report is a section on exactly the point you're talking about, public hospitals. They are absolutely a part of this whole emergency response team. They have to have not only surge capability, but they have to have communication capability with people that they've never had the necessity to have that kind of communication with before.

My suggestion would be that you have to get some people who are really skilled in the area of communications to answer that question. I'm not in a position to answer it. I mean, if you get a major attack on a city that takes out cellular telephone facilities and relay facilities for pagers, you have a serious problem responding.

Now, there are communities that have a plan that in the event of an emergency, emergency workers at hospitals are told if there is an emergency and you hear about it, here's what you do. So they don't have to be communicated with. It's kind of a standard operating procedure for what you do if something terrible happens downtown, and I think that is something that most good hospitals probably already have in place. I don't know that for a fact.

Mr. MURPHY. Well, I know some of the ones that I used to work at are doing that very thing. But they had also hoped to recognize that sometimes if there is what's a biological or chemical attack, obviously there are chemical issues, and they also have to be told how to get there and other routes to take and how to not be caught up.

But one of the things we learned from September 11 is that so many people were trying to use pagers and cell phones that the system shut down. I know around here they thought, well, BlackBerries worked, but now we know that if you overload that system, that shuts down too. And so we do have that other issue. And I am wondering in terms of other funding aspects if there's any direction we should be looking at here to help with the communication network.

Senator RUDMAN. I think when you have to—when you set national minimum standards, that will be part of the issue: What are the minimum communication requirements for a reliable system under adverse circumstances? Nobody seems to know that yet. We've got—now here in D.C., there is an interesting experiment going on with the use of different kinds of software that means you don't have to replace all of the equipment. And that's a very interesting approach. It's being done here and I'm sure there are people in the D.C. Police Department and the Park Service that can tell you about what they're doing.

Mr. MURPHY. Dr. Smithson, do you have any comments on those issues?

Dr. SMITHSON. Yes. In the aftermath of Aum Shimrikyo's attack in the Tokyo subway on March 20, 1995, physicians at St. Luke's Hospital, which received a flood of patients within minutes of this event, were reduced to shouting down the halls at each other in order to communicate.

New York City on September 11th and some of the other things that we've seen transpire tell us that we've got a communications issue. Having traveled so many places and talked with so many jurisdictions, I can, unfortunately, tell you that there are only a few that are making real progress on regional hospital planning. This is a very, very tough nut to crack because of the privatization of U.S. health care.

And that is why Federal assistance in this regard will be so crucial. We need to get hospitals in the same room with each other to plan and we need to develop plans to reach out to these medical health care providers.

And if communications go kerflooey, do you know what will save the day? People will be well trained if we have standards. They will know where to go when disaster strikes and what to do. Unfortunately, absent standards and absent sufficient funds for regional hospital planning, there are only a couple of places in this country that I know that have a shot at that right now.

Mr. MURPHY. In the few seconds that I have left, one other issue that hospitals have brought to my attention is with HCFA confidentiality standards. They wonder, will there be allowance to let a lot of it be thrown out the window in the case of an emergency and they don't have to jump through hoops to protect all kinds of things that they are trying to get information quickly across whatever methods that they have.

Dr. SMITHSON. Actually, I will use New York City as an example here. They have made tremendous strides with something called syndromic surveillance, which I think is rather misunderstood. This has great potential to provide us with leading-edge indication that there is a disease outbreak in communities.

And one of the things that they and others are working on is how you get crucial information without revealing sensitive information. I think they have made some good strides there.

Mr. MURPHY. Thank you.

Mr. SHAYS. Thank you.

Congressman Ruppersberger.

Mr. RUPPERSBERGER. I am probably going to make more statements than a question, but I will get to a question.

First thing, I think the premise is, if we cannot protect our first responders and give them the resources to do the job, then they can't protect us.

Now, Senator, your opening statement and Dr. Smithson's were excellent. I think—let's get to the bottom line as quickly as we can, and hopefully this won't be another hearing that we have and we all waste our time and nothing occurs.

The bottom line is this, that we have not made this a priority—we have not made homeland security. I would not want Governor Ridge's job. I mean, Governor Ridge is trying to take Customs and immigration officers and make them sky marshals, two entirely different disciplines.

Why does he have to do that? Because he is not getting the resources.

The administration—believe me, this is not partisan. Many friends of mine on both sides of the aisle are very much concerned about the priorities of moneys and resources going to first responders, and that has to be dealt with.

And I don't want to get, because it is really a side issue, but the fiscal policy right now is not working. You alluded to it, whatever it is, whether you are in favor of the tax cut, you are not in favor. But we have to reevaluate. If we are going to be out there taking the offensive on terrorism, dealing with homeland security, dealing with issues, Iraq, Iran, North Korea, we have to have the money.

And this deficit is looming out there. And as long as it is there, I think the administration is going to be extremely reluctant—they have showed it already—not to give the resources.

Now, if you are going to give somebody money, you have to hold them accountable for it. So the bottom line is that I think that your assessment for a national program is great. However, it is going to be very difficult, in my opinion. You can't tell me that—I described my district in the opening statement with an airport, a port and Army bases, versus in the middle of Utah, which I have traveled, and there are a lot of rocks out there. But you can't have both. And so it has to be, I think, looked at from a regional perspective and looked at where the priorities are.

Second, you have to look at who is doing it well now. Now, I know in the State of Maryland, I happen to be vice chairman of our shock trauma system, which is a regional—rated one of the top systems in the world. It is very good. I don't mean to be parochial, but there is a system that works. They have had a lot of training exercises. It works from the training of the paramedics to the communication, all over.

And there are other areas in the country that have this too. We need to take the ones that work well and look at what we are doing, so we don't waste a lot of money in training, in developing where we need to go.

Now, from—the general question, because I have to get to a question now, the question basically is, where do you start? And I think we have to be—from a public relations point of view, so the administration will listen and give us the same resources that Rumsfeld is getting, because, you know, sooner or later, as you say—I am on the Intelligence Committee. I know where we are. I can't tell you where we are, but I know that there are a lot of issues out there.

Al Qaeda is for real. Terrorism is for real. They are very patient. We have to be ready. Right now, you don't even have to be a terrorist if you want to go into a subway situation. And then there needs to be a response to that. That takes communication, it takes medical, it takes first responders, all of this coming together.

So my broad question is, how do we get the attention of the administration to reprioritize where the resources need to go? Because that is bottom line.

Senator RUDMAN. I have a very straight and simple answer to that, Congressman, that is, that you have to have a crash program mandated by Congress. There is a lot of support for it in both Houses, to set national minimum standards for first responders.

Because, until you do that, you cannot go to the administration, you cannot go to the appropriators and, with any sense, say, we need X amount of money. Because then you are going to get the same answer that you have gotten over the years on many other subjects: Well, there is going to be waste and fraud and mismanagement.

But if you can point specifically to, here are the national minimum standards, and you take the fire, police, emergency responders of 100 of the largest cities in America and say, here is the standard and here is what they have, then the message starts to sink in.

Mr. RUPPERSBERGER. Let's get down to the basics.

Mr. SHAYS. I hate to interrupt the gentleman. I will give him more time.

Dr. Smithson, do you agree with that?

Dr. SMITHSON. I do agree with that. Pardon me for being a bit uppity here, but you guys write the laws and you control the purse strings.

Mr. RUPPERSBERGER. That is what my next question is going to be. You are talking about Ridge—well, I am in the minority, though, and I am a freshman. So I have a chairman who is very strong.

Senator RUDMAN. That is not a disability, Congressman.

Mr. RUPPERSBERGER. Getting back to the issue, Ridge. You have made a recommendation. How would you implement that recommendation?

Senator RUDMAN. I would pass legislation, and there are people in the Senate who are willing and able to do that. I would make this a key issue. I would raise it—elevate it to high visibility, and I would get legislation on the books which mandates the administration by a date certain to establish national minimum standards for first responders for homeland security.

Mr. RUPPERSBERGER. Well, if you are going to do that, don't you think you should have a recommendation on what those standards should be?

Senator RUDMAN. No, I don't think you can do that. Because I don't think anybody truly knows. The expertise to establish those standards is out there in the country amongst your constituents, and together with Federal officials they can establish those standards.

I mean, we have standards now. I mean, you talk to the International Association of Fire Fighters, talk to the insurance companies. There are standards for fire departments, depending on the kind of a city you are in, for fighting fires.

There have to be standards for homeland security first responders across the board. Until do you that, you can ask for money until the cows come home, Congressman, and you are not going to get that.

Mr. RUPPERSBERGER. You know it.

Another thing that is very disturbing now, and we have had this in other committees: Because of terrorism, there are so many of the resources that the locals are taking away from other areas. There is another issue out there, which we are going to have to get on top of, that so much money is being taken from drugs; and that is still is our No. 1 threat. And so, you know, this is an issue that has to be dealt with.

I agree with you. I am sure Chairman Shays is a very forthright, aggressive person. Maybe we can develop something to that effect.

Senator RUDMAN. I want to make one response to what you said. And I think you will understand this very well, knowing Baltimore and Baltimore Airport.

The attack took place with airplanes. What was our response? To pour an enormous amount of resources into airports. I don't know how many billions it is now. But we have done all of this with airports. Now, maybe that was the right thing to do. But you always have to ask yourself the question, was that prioritization or was that responding to what we saw as an issue? Maybe a little bit of both.

Mr. RUPPERSBERGER. Well, there are a lot of ridiculous stories.

Mr. Chairman, the Port of Baltimore that I deal with, a lot of issues we have in port security. Baltimore happens to be a ro-ro, which is a lot of vehicles coming on and off. Because they are not a container port, 50 percent are ro-ro; we are not getting as much money as the container ports.

I mean, it is just—that is what I am talking about. I am very much concerned about the standards. It is going to be very difficult to pull these standards together.

Senator RUDMAN. But it can be done.

Mr. RUPPERSBERGER. Thank you.

Mr. SHAYS. Thank the gentleman.

Mr. Tierney.

Mr. TIERNEY. Thank you. I am not as nice as my colleagues, unfortunately, and I guess I don't really mean this to be partisan, but I do mean it to be a difference between the House and the Senate and the White House.

There is—there is policy, and then there is execution, and whether it is a Republican or a Democrat in the White House—and I hope we can see beyond that or whatever—there has been no execution.

Back in October 2001, there was a bipartisan letter that went to this administration, to Mr. Ridge and to the White House, asking them for a threat assessment. October 2001. There have been calls made, there have been individual letters sent. We have been haranguing and harassing this White House and this administration for a threat assessment since the very outset, after September 11th.

And I think that your recommendations are excellent, both of you have made them for some time now. I think that members of this committee, both Republicans and Democrats, have heard them. This chairman has been excellent on this issue; he has been a leader in the true sense of the word on this issue, and not hesitant to get on board a bipartisan effort; and he sees beyond the idea of whether or not this is political and understands the gravity of the situation in this country.

We simply have to, as a Congress, get beyond that. We have to sit here and decide whether or not we are going to do what the Senator recommends, and think Dr. Smithson also agrees with that recommendation, that we have to raise the level, put up the temperature here and demand some execution.

We should have had this threat assessment by this time. And I note that one of your recommendations, Senator, is, by the end of 2007 we have a threat assessment like that. That would be great, but that would be too late. It is probably the best that we can do in this scenario, but it is an unbelievable disgrace that despite the efforts of this Congress, this administration has failed to execute on that level. And we have just got to increase the pressure and work together on this.

And I know it is going to be more painful for my colleagues on the Republican side than it is for us, because of the party that is in the White House, but this is beyond that, as you have said, eloquently; and I think we all agree here, we have to demand those standards.

There are people out there that can do it. We can start by just asking the two of you to identify people; in about 5 minutes we would a list well worth going to. You have RAND, you have other institutions out there that do great work in these areas. There is no excuse for not having it done.

I think that if we look at just some of the recommendations in your testimony, and this Congress can do it, this committee can start doing it, by demanding that DHS and HHS put in their next budget a detailed methodology of how they are going to determine—it, in answer to my colleague's question, is probably best that they set forth a methodology of how we are going to get this information, then go about getting it and have a time line on that.

I think we do have to have a system that is more transparent and fairer about how we are going to distribute these moneys, and we can do legislation along that line, or put some constraints in our budget, requiring that these things be done.

Also, the most important one that I see in terms of my local responders here is some adamant statement about a distribution timeframe. You know, the next time that we put an appropriations bill out, I know we can do that. So I don't have a question here, but I don't think it is necessary. You have had an eloquent discussion of the issues here, and the recommendations from both of you individuals, I thank you very sincerely for laying it out so plainly and making it clear that these things are not just on the table today; they have been on the table for some time, and it has been a case of inactivity.

And while I think that this committee and Congress have acted, I don't think we have acted forcefully enough. And I think it is time to set aside the timidity and set aside the partisan aspects of it, and recognize this as simply a prerogative of Congress to kick the White House in the backside when it is necessary to get it to execute on these things.

I hope that we can work together to do that, to get the assessment, to get some timeframes on when distribution is going to get out there, and to get a real fair formula for that distribution and get moving. So I am happy to work with my colleagues on that.

I want to thank both of the witnesses again for their excellent testimony and all of your service to this country.

Mr. SHAYS. I thank the gentleman for his comments. And just—this may seem a little off subject, but it really relates to what you have said, Mr. Tierney. I think that the challenge for the majority party, frankly, is to be demanding a little more accountability on the part of the administration.

And we have done a disservice to the administration, because the number we are seeing for our needs in Iraq—if we had asked sooner and demanded a number sooner, I think they would have been better prepared today than they are. And I think that it does now also relate to your whole issue of what the threat is.

Before we had reorganization, Mr. Tierney was saying, don't we need to know the threat before we reorganize our government? And I felt we could do them in tandem. But I did feel we needed them, and I joined him in his desire to learn that. But we still don't have a threat assessment, which is, to me, pretty astounding, given that all three commissions said, know the threat in specific ways, then design your strategy around the threat, and then reorganize.

What I want to ask you about is that my first reaction when you came out with your number was that it seemed to be a contradiction; and you really answered it, which I wanted to state. And that is, if we don't have a way to evaluate what we need, how can we then come up with a number that says we need this?

What it appears is that your number was really an expression of what was requested on the part of local communities and States, and that it was a real estimate, but it was also kind of, I view, as a wake-up call. In other words, we need something far more than what we have now, and that is how I reacted to it.

Is that the way I should have?

Senator RUDMAN. Not quite. Almost. We got numbers from virtually all of those originations. Curiously, although they tried very hard, the police chiefs association could not give us—they just weren't comfortable giving us a number.

The fire gave us a number. We did not take these numbers as they were submitted.

Mr. SHAYS. Let me just say, parenthetically, we had a hard time getting the police departments to come. I asked, why are we just hearing from fire? And the police departments and the people we contacted just didn't want to step forward at this hearing and deal with this issue.

Senator RUDMAN. I have no explanation for that. We certainly had their cooperation. We couldn't get a number from them.

The fire departments across the country, we got good numbers from them and the responders. We adjusted those numbers somewhat. We did not just take the numbers we were given.

Now, the other thing that was very difficult, and I am sure the chairman will understood this was to get a number that truly reflected what the delta is of expenditure by city, State, counties on terrorism-related expenses versus their normal budgets. So, for instance, you take the Arlington fire department—and you are going to hear from the chief of the D.C. department—what is their foundation budget, and what is it that they are spending on homeland security issues, first responder preparation, above that?

We did a lot of work in that area. That is why if you look at the report you will find we worked with two of the best budget organizations in the city to help us put the numbers together, private organizations that are budget related. We therefore established a range, which is why in my opening remarks I said that we cannot tell you an exact number; but you can find out that number if you have a threat assessment and then mirror against that threat assessment what national minimum standards are for first responders. Then you will have the number.

Mr. SHAYS. And I am going to ask Dr. Smithson to respond to the fire-police issue.

But what I am struck with is, there may be a reluctance on the part of the administration to do this because it may set a standard that they don't think they can reach.

But wouldn't you agree that even if we knew that we needed to spend a certain amount, and we, setting priorities, thought we couldn't reach it, that would be better than doing what we are doing right now?

Senator RUDMAN. We say, set the national minimum standard, and then decide how many years it is going to take you to get to that standard.

And, most importantly, when you decide that, the decisions on where to start first cannot be made politically. Every congressional district gets a pro rata share. You take a look at where the threat assessment is, and you make sure that is where the money goes first.

Obviously, if I were king and you asked me, what would you do first, I would make sure that the 100 largest cities in America, the 100 largest cities in America, had chem-bio equipment for their first responders and first-rate communications and surge capacities for their hospitals.

If you would give me my priorities, based on working on this now for 7 years, since I have been involved in this issue, that is where I think the money should go. But can I tell you how much? No, I cannot tell you how much that would cost. But that is a question of multiplication, once you decide how much people need it.

Mr. SHAYS. When I was just making reference to the police and the fire, there are two different cultures here, you started to appear like you were going to respond to it.

I would be curious to know what your comment is.

Dr. SMITHSON. One of the reasons why training has not propagated as widely among the police is, basically, that their workday differs from those in the fire service.

Fire service, you know, when the bell rings they go out. But a considerable amount of their time is spent in the station. For police officers, they are in the station for a brief period of time for roll call and morning news or shift news, and then they are out on patrol.

So it has been much easier to get training propagated to the fire services. That is why I kind of, you know, made a face about the cultural differences there.

Listen, we are all waiting for this threat assessment, But in the absence of a threat assessment—and I don't know when it is going to appear—what we need to decide to do is to do the most good for the most people, and that means going by population density. Institutionalizing standards will bring the rest of the boats up with the tide.

And as for the recommendation that you go with the 100 largest cities, fine, but let's define "city," because when the domestic preparedness program was implemented, they went to the 120 largest metropolitan areas; and that meant the same training trooped to the same locations three different times.

Mr. SHAYS. I don't know what you just meant. What do you mean the same training went to the same——

Dr. SMITHSON. The same training went to the larger Denver area three times; it went to three different jurisdictions in the larger Denver area three times.

What we need to do is to get the Denver area—sorry to make an example out of them—working regionally. And that means putting—making some type of a calculation to get the funds working on a regional basis.

I know that is going to be tough, but if we think about this, we can get this done. And therein you will reinforce the mutual aid system; people will know who to call when it hits the fan, instead of wondering where the help is next door, and what it is that they can do to help.

So that is why a regional approach is really advisable here.

Mr. SHAYS. Let me just say I would be happy to give another round to whoever wants it.

Mr. TIERNEY. If I might, I just want to make—I think we have some responsibility to deal with the unease of our citizenry. And the way to deal with it is to plan, is to have the things that we are talking about.

It is just incredible that since September 11th our population, our constituents, from the fire fighters and the police officers all of the way down to the average person, regular person on the street, are in less than a comfortable situation. They sense great unease, because no one is telling them what the plan is.

I think they can deal with it if we haven't got to the end of the plan yet, if we are moving in that direction, if—the uncertainty and the great feeling of unease come from the fact that, as the government, we have yet to give them an outline that we know, what we can identify as the problem; and we have to plan to get there, to some solution in a reasonable period of time.

I think that is our charge. And thank you again.

Mr. SHAYS. Any comment on that? Any other member?

Mr. RUPPERSBERGER. Yes.

First, I think the regional approach is right. Again, we are asking for limited funding. We need to be held accountable for the results.

And the regional approach is not duplication of effort; and if you want to have money, a requirement should be that the region comes together with a plan.

Most of your counties and regional areas are larger than the cities in the middle. But the cities are the focal point. Because you talk about—again, I will be parochial to Baltimore. There are 2 million people in the Baltimore metropolitan area. Baltimore City has 630,000 people. But you just say "Baltimore." It has to be a region.

But that needs to be part of a plan. That is why I am concerned. I agree with everything that you said, and hopefully we will get something out of this hearing. I know that the chairman is a very proactive person, and he will run with this.

But, bottom line, it is still about what type of plan are you going to put together and then get the votes for? I mean, we might need a vote of somebody out in Utah. But we need the vote of other areas.

And you are saying basically, you do the regional approach where you can help the most people, and then you will bring it up the other way through the standards, the national standards; and that is how you will justify getting the money, correct?

Dr. SMITHSON. Yes.

Mr. SHAYS. Just before we have you all go, the other area that I wrestle with, I totally buy into your point about a crash program; and I know, Dr. Smithson, you do as well. But I have a hard time understanding how we allocate Federal, State and local. There are certain things that—frankly, I have my fire and police come to me, and I say, with all due respect, that is Bridgeport's responsibility or that is the State's responsibility, and don't use terrorist issues to try to get that money.

How do we deal with that issue?

Senator RUDMAN. Well, this is a personal opinion having been in State government in New Hampshire for many years. I am not a great fan of money going to States. I just never have been. It is another layer of bureaucracy.

Mr. SHAYS. You want it to go straight to local?

Senator RUDMAN. Yes.

If you are talking about the—let's say the Washington, DC, region, whatever that means, Washington metropolitan area, and I agree it shouldn't be the 100 largest cities, the 100 metropolitan areas, find those areas and decide what the standards are, and make sure that every police, fire and emergency responder organization starts to meet those standards with the money that is available.

I don't think you need another State planning organization. You start getting competition with Governors, whom I respect a great deal, but that is not their issue. It is not the State that responds to this; it is the local governments that respond to this. And so I truly think, if you start looking at State grants—now, you may need State grants for certain activities, but if you talk about what we are talking about, the first responders don't work for the State, they work for communities.

Now, that is my opinion. Others are free to disagree.

Dr. SMITHSON. On this issue of burden-sharing, I couldn't agree with you more, which is why I made it a focus of Chapter 7 in Ataxia.

I have yet to meet the politician——

Mr. SHAYS. Chapter what? I'm sorry.

Dr. SMITHSON. Chapter 7 in Ataxia. You will see a series of proposals in there for how burdens can be shared between the Federal Government, the State government, and local jurisdictions. Who is going to pay the bill here is a big part of this discussion. And if we don't get that straightened out soon, we will see additional backsliding in the preparedness gains made to date.

Mr. SHAYS. Right.

Dr. SMITHSON. I have yet to see a politician who likes to talk about higher budgets or new taxes. But there are formulas that are available out there. And I agree with you, we should be going with the models that work.

In this particular case, locally, I will urge you to look at what has happened in Florida, which passed a disaster fund bill, and how at the State level, this is done with a tax. They have local moneys going into disaster preparedness. And they are getting better prepared with each passing year that this is involved.

There is definitely a part of this burden that the Federal Government should carry. But locals will be the first to tell you that all emergencies are local, and they need to carry their part of the burden.

Mr. SHAYS. Great point. Let me just ask, in this national minimum standards—when we do that, would we also allocate to Federal, State and local how we think the resources—where they should come from?

Senator RUDMAN. I think that it is a wholly different process altogether. I think, first, you set the standard, then policymakers have to decide how you are going to prioritize the distribution of whatever money is available against that standard.

Mr. SHAYS. OK.

Mr. Murphy.

Mr. MURPHY. Mr. Chairman, I just want to echo Senator Rudman's comments on the way things work.

I know with fire companies in my district and other first responders who have received some Federal grants, oftentimes they sit around scratching their heads, trying to guess what is going to be funded. And it appears to me that the self-contained breathing apparatus and radios seem to be the magic thing. Yet many of them say, who are these faceless, unnamed people in some Federal bureaucracy somewhere deciding what we need, when we know what we need in our communities? Then they have to guess.

And those who are lucky enough to guess that they needed some breathing apparatus got it; and those who already had all of those things, but needed something else to help with response, didn't. It is an absurd way to run things. I think it works much better if there are established community standards.

Knowing that happens in my district, which probably happens in many other districts around America, is there are so many fire departments, and every town and hamlet wants their own police department, police chief, as everyone else—that each one wants a full complement of equipment for everything. When, if they really look at what was needed regionally—for example, I have a strip of highway, 3 or 4 miles long, that I think there are seven fire departments within that area. Each one has to have their own everything. And getting them to coordinate that effort is very difficult, and much better if there is a standard. Say, here is a standard; if someone within this stretch needs that equipment, it must be shared.

If we don't set the example on the Federal level, then it kicks back to the States. The States won't take a stand on it either.

So I am in absolute agreement that we need to do that.

Senator RUDMAN. I want to comment.

That is absolutely right, and if you look at the cell, you will find there were certain types of equipment that you don't need every department to have, certain types of chemical detection equipment. As long as you decide how-many-mile radius it is, then it can be deployed.

Mr. TIERNEY. I was wondering, Mr. Chairman, if Dr. Smithson could make available a copy of her report for the committee?

Dr. SMITHSON. I would be delighted to. It has also been up on the Web since October 2000.

Mr. TIERNEY. At?

Dr. SMITHSON. Www.stimson.org. I have it on CD-ROMs. More than happy to make it available.

Thank you, sir.

Mr. TIERNEY. Thank you. And you came by the name of that document how?

Dr. SMITHSON. Ataxia. It means confusion, and I think there is a great deal of confusion about the nature of the terrorist threat. It means lack of organization. And, well, I would argue that we are still not as organized as we could be. And last, but not least, it means involuntary muscle spasm. Since the research was principally about chemical and biological response, well, you can see where that one came from, too.

Mr. SHAYS. I wondered that, too.

Mr. TIERNEY. Thank you.

Mr. RUPPERSBERGER. We have had a lot of hearings, but I want to congratulate both of you. You have been focused, forthright, and you have come to recommendations that we can move forward with.

I was in local government for over 18 years. I can tell you whether or not—it is not partisan, but as a county executive, when I got money from the COPS Program, as an example, it came directly to us right away—putting cops on the street right away.

Anything else that, when you had to go through the State, the Federal bureaucracy, the State bureaucracy, half the money is gone before you ever got it. Now, right now, we have a program, that fire fighters grant program, it is the most popular program right now, where the money goes directly to the fire departments. Why can't we use that program now, as a role model, to move forward? It is a direct program. It is just like the COPS Program was.

They are changing it?

Mr. TIERNEY. They are moving it out.

Mr. RUPPERSBERGER. Here is an example of—we—I think there is agreement on this panel that we have to do something. And if we don't do something, something is going to happen, then we are going to be scurrying all over the place. But at least—and you can't conquer Rome overnight.

The only thing I am concerned about, I keep getting back to this national program. If you are going to accomplish something, it can't be broad; you have to be focused with it.

I am just concerned that if we come out with a national program, just say, here it is, we have to kind of, I think, tailor it from a recommendation point of view. If we are going to try to get votes from our colleagues, I just think we need to have a little more specificity.

Can you help me there at all?

Senator RUDMAN. I don't think there ought to be a lot of partisan debate about setting, A, a threat assessment, or, B, national standards. Now, once you have that done, then you are going to have some discussions about distribution.

Mr. RUPPERSBERGER. How long is it going to take to do that?

Senator RUDMAN. I don't think all that long if you want to get the right people to do that.

Mr. RUPPERSBERGER. You can put in the bill a timeframe.

Mr. SHAYS. But the gentleman—but what Mr. Rudman is saying, let's just know the reality; then we will deal with the politics after that.

Senator RUDMAN. Nobody knows the reality right now. No one.

Mrs. MALONEY. But once we get to the politics, and we are dealing with the politics now in the high-threat areas, and if we direct to it 100 regions, we don't have the votes to pass it, because my colleagues will say, I live in a desert someplace, and we have—we are afraid of desert rats or something.

Senator RUDMAN. Congresswoman Maloney, let me just respond by quoting Winston Churchill, when he said that democracy was the worst form of government, except for all of the others that have been tried.

Mrs. MALONEY. Very finally, your comment on your response to the airports, where we went in and made that the top priority, and it may not have needed to be the top priority. Can you elaborate? Do you think we made a mistake in putting so many resources in one area?

Senator RUDMAN. My own personal opinion is that some of that money should have gone to port security, because I think that is a larger threat to major metropolitan areas.

No question, we had to do a lot with the airports. But when you look at the number of billions spent, it was a crash program; there were no standards really thought about. It was done, it was hurried through Congress, there were fights about whether there would be privatization of the force and so forth or ended up public.

My own personal opinion, having looked at this for a long time, is that some of that money would have been better spent in port security on America's major East Coast, West Coast and Gulf ports.

Mr. SHAYS. Let me just quickly ask, because this committee is going to act on this, the national minimum standards that we would ask to be set up, are we asking DHS to do that?

Senator RUDMAN. Yes, you are. They have—right within FEMA, they have the capability to do it. And they have all of these groups around the country that they have liaison with, who are more than willing to sit down and help.

Mr. SHAYS. Time frame we should give them is how long?

Senator RUDMAN. I would not give them probably more than 9 months.

Mr. SHAYS. OK. We do appreciate both of you being here. Thank you.

Our second and final panel, that is, Adrian H. Thompson, who is the chief of D.C. Fire and EMS Department in the District of Columbia; and Mr. Edward Plaugher, fire chief, county of Arlington, VA, International Association of Fire Chiefs.

Does that mean you are the head of it or are you just part of it?

Chief PLAUGHER. Part of it.

Mr. SHAYS. We are also joined by Deputy Chief Sellitto, who is with Chief Thompson. We will have two statements, but all three of you, we will swear you. If you would stand and raise your right hands, please.

[Witnesses sworn.]

Mr. SHAYS. Note for the record, our witnesses have responded in the affirmative.

We will start with you, Chief Thompson, and appreciate that both of you are here. I would love to have you be able to incorporate any part of your statement into what you have already heard that you feel inclined to make, as well, in your statement.

And we will do a 5-minute, and then we roll it over 5 minutes.

## STATEMENT OF ADRIAN H. THOMPSON, CHIEF, DISTRICT OF COLUMBIA FIRE AND EMERGENCY MEDICAL SERVICES DEPARTMENT

Chief THOMPSON. Thank you. First, I would like to say good morning, Chairman Shays, members of the Subcommittee on National Security, Emerging Threats and International Relations. I am honored to be here to present today.

I am Adrian H. Thompson, chief of the Washington, DC, Fire and EMS Department. With me is Michael Sellitto, deputy fire chief of special operations, and my acting assistant fire chief for operations, Chief Doug Smith.

The District of Columbia Fire and EMS Department has been participating in the State Domestic Preparedness Program administered through the Department of Justice, Office of Domestic Preparedness, and is continuing to receive funding through this source. Although there is a delay from submission to receipt of funds, the appropriated funding eventually becomes available. We are also making use of Fire Act grant programs and are currently awaiting a decision on a possible award.

The Department also submitted requests to the Federal Government for financial assistance in 2001, immediately following the terrorist attacks of September 11th. Specifically, we requested support in those critical areas where we judged we might have a shortfall if an event were to occur which caused loss of personnel or response equipment.

We made requests for additional fully equipped fire apparatus, because having a "ready reserve" fleet is essential for sustained response during a major event or multiple-site incident that would otherwise strip the remainder of the city of essential services. The request also included additional technical rescue and hazardous material equipment, which would allow us to operate in the most effective manner. The funds were appropriated and expanded in accordance with our request.

One important issue is the development of nationwide equipment standards so that emergency equipment can be shared across jurisdictional lines. Depending on finalized standards, the equipment bought earlier may have to be replaced to meet them. In addition, with the purchase of many highly technical pieces of equipment comes the added maintenance cost of this new equipment. It is hard to project necessary operating budget funding for this purpose, especially since much of this equipment was obtained after the fiscal year budget was finalized.

Another major area in which we saw a need was an inadequate number of personnel with the necessary specialized training to perform the tasks involved in technical rescue and hazardous materials incidents. To fulfill these needs, we made use of the many Federal programs, including: Department of Homeland Security, Office of Domestic Preparedness programs; FEMA, Emergency

Management Institute programs; and Fire Administration, National Fire Academy programs.

These programs are all worthy of continued funding, as they are the standard by which all other training is measured. All of the above offer free training to agencies such as ours. However, this agency incurs overtime costs whenever we send members to training. Since our agency operates on a 24-hour 7-day-a-week schedule, overtime is necessary to maintain adequate staffing while members are in training. Our September 2001 request asked for specific language permitting use of appropriated funds to cover this expense. This request was honored, and as a result, we have achieved a major accomplishment in the area of specialized training. We now have a greater number of trained individuals, and in many cases, the individuals have been trained to higher levels than they were capable of in the past.

Unfortunately, government training programs do not cover all areas of technical training, and it was necessary to contract out with private and institutional vendors for some of this training. The increase in numbers of personnel with specialized training will result in a corresponding increase in recertification costs to maintain the numbers at the new levels in the future.

All training and procedures must be practiced regularly to ensure readiness. Training drills and exercises are now multiagency events. There should be a mechanism to encourage participation from local agencies which serve a support function during emergencies. These support agencies have limited funds to participate in these exercises, and as a result, full participation is not always possible.

In regards to some of the shortfalls that we foresee in our agency in the city, in terms of equipment needs, to keep the latest technology available in areas of emergency medical services and biological and chemical field testing, continued funding for upgrades should be provided.

To address our equipment maintenance needs, consideration should be given to allowing a percentage of our future funding to be allocated to equipment maintenance and upkeep.

To address our training needs, it is essential that training requests be funded with specific language allowing us to use the funds for overtime purposes for backfill positions. In addition, first responder agencies should be allowed to contract with the private sector to fulfill needs unmet by the Federal programs and to use the training funding to support recertification as necessary.

Finally, funding for training exercises must be available to pay the overtime costs for support organizations to ensure their participation in an exercise.

I must point out the District of Columbia is unique, in that appropriated funds do not have to go through multiple layers of State, county, and local government to get to the point where the first responders can use appropriated funds. Unfortunately, our colleagues in surrounding jurisdictions have told us that funding no longer comes directly to them, as in the past, and must filter down to them through the middle layers of government, thus delaying the receipt of funds as well as decreasing the ultimate amount that they may receive at first responder levels.

I would like to thank you for this opportunity to present our views, and am prepared to answer any questions you may have.

Mr. SHAYS. Thank you, Chief Thompson. We will have questions.

[The prepared statement of Chief Thompson follows:]

GOVERNMENT OF THE DISTRICT OF COLUMBIA
FIRE AND EMERGENCY MEDICAL SERVICES DEPARTMENT
WASHINGTON, D.C. 20001

★ ★ ★

**Combating Terrorism: Preparing and Funding First Responders**

**Testimony of Adrian H. Thompson, Chief, District of Columbia Fire and Emergency Medical Services Department**

**Subcommittee on National Security, Emerging Threats, and International Relations, Committee on Government Reform**
**U.S. House of Representatives**
**Room 2247 Rayburn Building**
**Washington, D.C.**

**September 9, 2003**

Good morning Chairman Shays and members of the Subcommittee on National Security, Emerging Threats, and International Relations. I am honored to be here to present testimony today. I am Adrian H. Thompson, Chief of the Washington, D.C. Fire & EMS Department. With me is Michael Sellitto, Deputy Fire Chief of Special Operations.

The District of Columbia Fire and EMS Department has been participating in the State Domestic Preparedness Program, administered through the Department of Justice, Office of Domestic Preparedness, and is continuing to receive funding through this source. Although there is a delay from submission to receipt of funds, the money eventually becomes available. We are also making use of the Fire Act grant program and are currently awaiting a decision on a possible award.

The Department also submitted requests to the federal government for financial assistance in 2001, immediately following the terrorist attacks of September 11.

Specifically, we requested support in those critical areas where we judged we might have a shortfall if an event were to occur which caused a loss of personnel or response equipment.

We made requests for additional fully equipped fire apparatus. Having a "ready reserve" fleet is essential for sustained response during a major event, or a multiple site event that would otherwise strip the remainder of the city of essential services. The request also included additional technical rescue and hazardous material equipment, which would allow us to operate in the most effective manner. The monies were appropriated and expended in accordance with our request.

One important issue is the development of nationwide equipment standards, so that emergency equipment can be shared across jurisdictional lines. Depending on finalized standards, equipment bought earlier may have to be replaced to meet them. In addition, with the purchase of many highly technical pieces of equipment comes the added maintenance costs associated with upkeep. As no historical maintenance costs are available with much of this new equipment, it is hard to project the necessary operating budget funding, especially since much of this equipment was obtained after the fiscal year budget was finalized. Another major area in which we saw a need was an inadequate number of personnel with the necessary specialized training to perform the tasks involved in technical rescue and hazardous materials incidents. To fulfill these training needs, we made use of the many excellent federal training programs, including;

- Department of Homeland Security, Office of Domestic Preparedness Programs
- Federal Emergency Management Agency, Emergency Management Institute Programs
- Fire Administration, National Fire Academy Programs

These programs are all worthy of continued funding, as they are the standard by which all other training is measured. All of the above offer free training to agencies such as ours. However, this agency incurs overtime costs whenever we send members to training. Since our agency operates on a 24 hour a day, 7 days a week schedule, overtime is necessary to maintain staffing levels while members are in training. Our September 2001 request asked for specific language permitting use of the appropriated monies to cover this expense. This request was honored, and as a result, we have achieved major accomplishments in the area of specialized training. We now have a greater number of trained individuals, and in many cases, the individuals have been trained to higher levels than we were capable of in the past.

Unfortunately, government training programs do not cover all areas of technical training and it was necessary to contract with private and institutional vendors for some of this training. The increase in numbers of personnel with specialized training will result in a corresponding increase in re-certification costs to maintain the numbers at the new levels in future years.

All training and procedures must be practiced regularly to ensure readiness. Training drills and exercises are now multi-agency events, and there should be a mechanism to encourage participation from local agencies which serve a support function during emergencies. These support agencies have limited funds to participate in these exercises, and as a result, full participation is not always possible.

**Potential shortfalls**

To address equipment needs:

***To keep the latest technology available in areas such as emergency medical services, and biological and chemical field testing, continuing funding for upgrades should be provided.***

To address equipment maintenance needs:

***Consideration should be given to allowing a percentage of future monies to be***

***allocated to equipment maintenance and upkeep.***

To address training needs:

***It is essential that training requests be funded with the specific language allowing us to use the funds for overtime expenses to backfill positions. In addition, first responder agencies should be allowed to contract with the private sector to fulfill needs unmet by the federal programs and to use training funding to support re-certification, as necessary. Finally, funding for training exercises must be available to pay for overtime for support organizations to ensure their participation in exercises.***

We must point out the District of Columbia is unique, in that appropriated monies do not have to go through multiple layers of state, county and local government to get to the point where the first responder can use the appropriated funds. Unfortunately, our colleagues in surrounding jurisdictions have told us that the monies no longer come directly to them, as in the past, and must filter down to them through the many layers of government, thus delaying the receipt of the funds as well as decreasing the ultimate amount they may receive at the first responder level.

Thank you for the opportunity to present our views, I am prepared to answer any questions you may have.

Mr. SHAYS. Mr. Plaugher.

## STATEMENT OF EDWARD PLAUGHER, CHIEF, ARLINGTON COUNTY, VA FIRE AND RESCUE DEPARTMENT

Chief PLAUGHER. Good morning, Mr. Chairman and members of the subcommittee. I am Edward Plaugher, chief of the Arlington County, VA Fire and Rescue Department.

I appear today on behalf of the International Association of Fire Chiefs, which represents the leadership and management of America's fire and emergency services. I am a member of the IAFC's Terrorism Committee and chairman of its Legislative Subcommittee.

I have submitted for the record a prepared statement from which I will highlight a few of the key areas during my opening remarks. The stated purpose of this hearing is to review the recent Council on Foreign Relations' report entitled, "Emergency Responders: Drastically Underfunded, Dangerously Unprepared."

The IAFC, the International Association of Fire Chiefs, participated in the development of that independent task force report and agrees with many of its recommendations. I will begin with a quotation taken directly from the Council on Foreign Relations task force report, found on page 11: "enhancing responder capabilities will require inputs on multiple levels. Providing response equipment is only one aspect of improving overall preparedness. Without appropriate staffing, training of personnel, and sustaining equipment and capabilities over time, new equipment may contribute only marginally to greater preparedness. Wherever possible, the all-hazards approach should be followed to ensure that, to the maximum extent possible, resources devoted to responding to a terrorist attack, can enhance underlying emergency preparedness capabilities for addressing natural disasters."

I continue, "With whatever capabilities they have, however, America's local emergency responders will always be the first to confront a terrorist incident and play the central role in managing its immediate consequence. Their efforts in the first minutes and hours following an attack will be critical to saving lives, reestablishing order, and preventing mass panic. The United States has a responsibility to provide them with the equipment, training and necessary resources to do their jobs safely and effectively."

The fire service—as was alluded to earlier today during this hearing, the fire service is the only entity that is locally situated, staffed, trained and equipped to respond to all types of emergencies. America's fire service is an all-hazards, all-risk response entity.

I can appreciate that the Federal Government's focus on terrorism is by preventing terrorist acts from occurring and enhancing the ability of emergency responders to mitigate an attack when it occurs. That is why maintaining and enhancing the current Assistance to Firefighters grant program, now in its 3rd year, is critically important to terrorism preparedness and response. And Chief Thompson was just talking about their application for the Fire Act, and they are waiting anxiously, as are a lot of us, for the results of that request.

Only when the baseline needs are met can departments enhance their capabilities. The Assistance of Firefighters grant program, we call it the Fire Act, is specifically tailored to accomplish this goal. Although funds in the Fire Act can be used to purchase specialized counterterrorism related equipment, the program is directed at addressing basic needs.

In the report, they found that approximately one-third of the fire fighters on a shift do not have self-contained breathing apparatus, and that 57,000 fire fighters across the country lacked turnout gear. It is important to note that all of these items are so basic to emergency response that in addition to enhancing a department's basic readiness, they will certainly be used in the event of a terrorist attack.

The IAFC strongly supports the Fire Act and urgently recommends that it be kept as a separate and distinct program under the U.S. Fire Administration. Under management by the USFA, a part of FEMA, the program has been an unqualified success. The reasons are, first, that the Fire Act grants are made directly to local jurisdictions after undergoing a competitive, peer-reviewed process which measures and ensures that money is being spent in a productive, responsible manner; second, the grants are needs-based, whereby the local fire departments must demonstrate an actual need for the proposed equipment or training; third, the grant program requires a copayment for the need from the local community by ensuring a "buying-in" from local officials for the specific equipment purchased through the funding; and, last, the law has a "maintenance of expenditures" provision which means that the Federal grant can only supplement, not supplant, local fire-fighting funds.

The important point to be noted is that the Federal funds are channeled directly to the fire departments for the purposes they are intended.

The IAFC also supports the terrorism preparedness block grants administered by the Office of Domestic Preparedness in the Department of Homeland Security. While this program plays an important role in enhancing response capabilities to certain elements of the first responder community, it has not proven effective for the fire service since we lack centralized representation at the State level. We have neither a State fire chief nor a secretary of fire safety within any of the 50 States.

Mr. Chairman, in closing, I would like to quote my friend, the late Chief Jack Fanning of the Fire Department, city of New York, who died at the World Trade Center on September 11, 2001. In testifying before a Senate committee earlier that year on terrorism preparedness, Chief Fanning said, "The emphasis must be placed on the most important aspects of the equation, the first responder and the first responder teams. If lives are to be saved and suffering reduced, it will be up to them to do it. At an incident, whatever the scale, fire fighters and other responders will be there within minutes, some quite possibly becoming victims themselves."

Chief Fanning's testimony ended with these words, "They, the first responders, will do what they have always done, act to protect the public they serve. Knowing this, let us provide them with the tools they need to perform their duties safely and effectively."

Mr. Chairman, the subject of your hearing is of great importance to America's fire and emergency service. There is no question that the Federal support is required. The key issue is effectiveness of that support. That is why we are solidly behind keeping and enhancing the current Assistance to Firefighters grant program as a separate and distinct program. The current program administrator, FEMA, has done an outstanding job which we, the IAFC, fully support.

I appreciate this opportunity to appear before the subcommittee, and will be pleased to respond to any questions.

And, Congressman Shays, I go back several years ago when we were also in a committee—subcommittee meeting prior to September 11, when we were talking about this very same issue, which was emergency responder preparedness. And the focus of that was a lot of what is contained in this report and what we needed to do at that time.

So, again, I am ready for your questions at any time.

[The prepared statement of Chief Plaugher follows:]

# Combating Terrorism: Preparing and Funding First Responders

Statement by

Chief Edward Plaugher

*presented to*

## Subcommittee on National Security, Emerging Threats, and International Relations

of the

Committee on Government Reform

## U.S. House of Representatives

September 9, 2003

International Association of Fire Chiefs
4025 Fair Ridge Drive • Fairfax, VA 22033
Tel: 703.273.0911 • Fax: 703.273.9363

Mr. Chairman and members of the subcommittee, I am Edward Plaugher, chief of the Arlington County (VA) Fire and Rescue Department. I appear today on behalf of the International Association of Fire Chiefs (IAFC) which represents the leadership and management of America's fire and emergency service. I am a member of the IAFC's Terrorism Committee and chairman of its legislative subcommittee.

The stated purpose of this hearing is to review the recent Council on Foreign Relations Report entitled: *Emergency Responders: Drastically Underfunded, Dangerously Unprepared.* The IAFC participated in the development of that independent task force report and agrees with many of its recommendations. Central to the findings of funding shortfalls for the fire and emergency service was the needs assessment prepared by the National Fire Protection Association. This report was a requirement by Congress when the Assistance to Firefighters grant program was enacted in 2000.

I begin my testimony with quotes taken directly from the Council on Foreign Relations task force report found on page eleven.

"Enhancing responder capabilities will require inputs on multiple levels. Providing response equipment is only one aspect of improving overall preparedness. Without appropriate staffing, training of personnel, and sustaining equipment and capabilities over time, new equipment may contribute only marginally to greater preparedness. Wherever possible, an all-hazards approach should be followed to ensure that, to the maximum extent possible, resources devoted to responding to a terrorist attack can enhance underlying emergency preparedness capabilities for addressing natural disasters.

"With whatever capabilities they have, however, America's local emergency responders will always be the first to confront a terrorist incident and will play the central role in managing its immediate consequences. Their efforts in the first minutes and hours following an attack will be critical to saving lives, reestablishing order, and preventing mass panic.

"The United States has a responsibility to provide them with the equipment, training, and other necessary resources to do their jobs safely and effectively."

Mr. Chairman, America's fire and emergency service reaches every community across the nation, covering urban, suburban and rural neighborhoods. Nearly 1.1 million men and women make up the fire service – 293,600 career firefighters and 784,700 volunteer firefighters serve in 30,000 career, volunteer and combination fire departments across the United States.

The fire service is the only entity that is locally situated, staffed, trained and equipped to respond to all types of emergencies. The fire department responds to natural disasters such as earthquakes, floods, tornadoes and hurricanes as well as to manmade catastrophes both accidental and deliberate. As such, America's fire service is an all-hazard, all-risk response entity.

The fire service responds to 17 million calls each year that include fire suppression, emergency medical services, hazardous materials, wildland fires and technical rescues such as high angle, swift water, building collapse, trench rescue, auto extrication and others. The fire service is also heavily involved in fire prevention and education, fire code enforcement and arson investigation. America's fire service protects the critical infrastructure of the United States and is part of that critical infrastructure – the continuance of government. Mr. Chairman, I would like to submit for the record a monograph prepared by the IAFC titled: *America's Fire Service* which gives a brief description of the numerous activities that define this emergency first responder community.

Mr. Chairman, in your invitation you asked witnesses to address the problems of matching appropriate funding amounts to an evolving and uncertain set of emergency responder requirements and to address the role of federal agencies in supporting local first responders and emergency management officials in the event of a terrorist incident.

I can appreciate the federal government's focus on terrorism – to prevent terrorist acts from occurring and then to prepare emergency response organizations to deal with them when they do occur. That is why maintaining and enhancing the current Assistance to Firefighters grant program – now in its third year – is so critically important to terrorism preparedness and response. Preparing fire departments to deal with specific acts of terrorism requires a fire department that is adequately staffed, equipped and trained to deal with the normal hazards and risks found in a specific community. Only when baseline needs are met can departments enhance their capabilities. And that is precisely why the Assistance to Firefighters grant program – we call it the FIRE Act – is so important toward that end.

The language that established the fire grant program states that the purpose of the program is for "protecting the health and safety of the public and firefighter personnel against fire and fire-related hazards." Although funds can be used to purchase counterterrorism-related equipment, the program is directed at addressing basic needs. And according to the survey recently completed by the U.S. Fire Administration (USFA) and National Fire Protection Association, many departments remain in need of basic equipment such as turnout gear, radio communications, and breathing apparatus that only the FIRE Act can address.

The IAFC strongly supports the FIRE Act and urgently recommends that it be kept as a separate and distinct program, under the U.S. Fire Administration (USFA). Under management by USFA, a part of FEMA, the program has been an unqualified success. The reasons for that are: (1) FIRE Act grants are made directly to local jurisdictions after undergoing a competitive, peer-reviewed process which ensures that the money is being spent in a productive, responsible manner, (2) grants are needs-based, whereby the local fire departments must demonstrate an actual need for the proposed equipment or training, (3) the grant program requires a co-payment for the need from the local community by ensuring "buy-in" from local officials for the specific programs and equipment purchased through the funding, and (4) the law has a "maintenance of expenditures" provision which means the federal grant can only supplement, not supplant, local firefighting funds.

The important point to be noted is that the federal funds are funneled directly to fire departments for the purposes intended. There is 100% accountability. Mr. Chairman, I would like to submit for the record the three-year history of FIRE Act grants delivered in the state of Connecticut as an example of this exemplary accountability.

The IAFC also supports the terrorism preparedness block grants administered by the Office for Domestic Preparedness. This is a formula grant program that runs through an additional layer of bureaucracy in the states. This method may well benefit some first response organizations but has not proven effective for the fire service since we lack centralized representation at the state level. We have neither state fire chiefs nor Secretaries of Fire Safety within any of the 50 states. Instead, the command structure originates from within local jurisdictions, with local fire chiefs occupying the top positions. That is why the FIRE Act remains the only effective federal program for assisting the fire service.

Mr. Chairman, in closing, I would like to quote my friend, the late Chief Jack Fanning of the Fire Department, City of New York, who died in the World Trade Center on September 11, 2001. He testified before a Senate committee earlier that year. In preparing for terrorism, "*the emphasis must be placed upon the most important aspect of the equation – the first responder and first responder teams... If lives are to be saved and suffering reduced it will be up to them to do it. At an incident, whatever the scale, firefighters and other responders will be there within minutes, some quite possibly becoming victims themselves.*" Chief Fanning's testimony ended with these words: "*They (the first responders) will do what they have always done – act to protect the public they serve. Knowing this, let us provide them with the tools they need to perform their duties safely and effectively.*"

Mr. Chairman, the subject of your hearing is of great importance to America's fire and emergency service. There is no question that federal support is required. The key issue is the effectiveness of that support. That is why we are solidly behind keeping and enhancing the current Assistance to Firefighters grant program as a separate and distinct program. The current program administrator – FEMA – has done an outstanding job which we fully support.

I appreciate the opportunity to appear before this subcommittee. I will be pleased to respond to any questions.

Mr. SHAYS. Thank you very much. We will start with the vice chairman of the committee.

Mr. TURNER. Thank you, Mr. Chairman.

The discussion that we have had here today has clearly identified huge holes in what our response has been since September 11th. And how to close it, it seems as an issue that—I am not quite certain, after having heard your testimony now in contrast to what we have heard before.

Having been a mayor, my view of communities is that they are separate and distinct and that the needs that they have are going to be both individual, based upon the experience, the investments that they have made to date; what equipment that they do have; and also the actual threat that specific community might have.

Also, their metropolitan areas are organized differently, and we had the discussion here, do we go through a major metropolitan area or do we go through cities. Some cities are very small geographically, but have very large metropolitan areas; some cities actually encompass their entire territory.

In talking about the issue of national standards, I mean, to recognize that there are some things obviously that we need to look at for national standards; and you mentioned some of those. Obviously, that goes to protective gear that first responders are going to have. That goes to not only saving their own lives, but also their ability to function and save others. So some of those are pretty easily identified as items that we need to undertake.

When you look at both then, States or a national threat assessment, I do get concerned as to how some of these issues might be lost. What are your thoughts—you heard the testimony in the first round—about a national threat assessment; and then some of the mechanisms that you have just discussed, a peer review and more local participation in identifying our threats?

Chief Thompson.

Mr. SELLITTO. We were lucky enough to participate in a pilot program with the Department of Homeland Security where we have just done a threat vulnerability, capabilities and needs assessment. It is part of a new computerized program that they have; and participating in the pilot program, I think that is the way they need to go.

Apparently, it is now expanding out to other States. And they are trying to capture a lot of that information that we were just talking about—the threat, the vulnerability, and everything else—to try to come up with a comprehensive package that I believe is going to lead to how we are going to allocate some of the additional funding in the future.

Mr. TURNER. So you did participate in a threat assessment program with the Department of Homeland Security?

Mr. SELLITTO. Correct.

Mr. TURNER. And it was tailored to your individual community? It was not—in other words, the process was to identify your specific threats and your specific needs rather than looking at the national level?

Mr. SELLITTO. It was looking at the National Capital Region. It was a regional effort.

Mr. TURNER. You thought the program was a good one that they had?

Mr. SELLITTO. It probably needs a little bit more development. Like I said, it was a pilot. But I think it was an excellent start; it is moving in the right direction.

Chief PLAUGHER. We also participated in that assessment, and we found several areas that needed improvement. It was a program that was—as the Chief was saying, that was brought—it was a computerized program that they asked the communities in the entire Washington metropolitan area to respond to various questions. Large numbers of actual elements were in the program.

What we found troubling about the system and the program is that prior to us taking that assessment, they didn't sit down and say, what would you think would be the key areas needed to be prepared for? They just kind of dumped it on you and said, here is a program, fill it out. And it was troubling from that standpoint. Because we said, what about these six or seven other key areas? And they said, well, it is not included in this assessment.

Well, that is kind of troubling, particularly since in your first do, and particularly my first do is the Pentagon, when we have a huge national target there. And why wasn't some of that focused into this assessment tool, and why weren't some of the other things that we thought critical?

So I think the chief is absolutely right. They need to go back and revisit the assessment tool.

They also need to get some direct local government input into that instrument, and also as part of that instrument allow for separation, national group and then a regional group.

Mr. TURNER. Do you know how the assessment was put together?

Chief PLAUGHER. They just bought it from a contractor and sent it out to us. And I have a real problem with the contractor too, but that is a whole other issue. It could be part of my bias. So I have to admit that.

Mr. TURNER. Thanks.

Mr. SHAYS. Mr. Tierney.

Mr. TIERNEY. Thank you, Mr. Chairman.

I would like to just keep it focused on one issue, because you have articulated most of the other, much to my satisfaction, and others I am sure.

The Fire Act and whether or not it continues to be administered the way it has been since its inception, or whether we allow it now to be changed so that the money no longer comes directly to the end user, I would like your comments on the wisdom or lack of wisdom in making the change; and what you think has worked about the program, having been funded the way it has been to date; and if you think it would be an improvement to go the other way, why.

Chief THOMPSON. In regards to the city itself and the fire grant appropriations that we apply for, I think our biggest concern is the copays that we have to pay. Our budget is limited, as it is. If you apply for a grant for $1,000, you have to come up with a matching or a copay with it. Your budget is tight as it is, when they are first drawn up; so if you have to apply for the fire fighter grant, you have to squeeze some money out of someplace else just to match the funds. It makes it kind of difficult on the rest of the year for the budgetary process in terms of making sure we are where we need to be just in terms of the money we allocated initially.

Also, streamline the turnaround time in terms of getting them back us to. We got a request in for a grant for PPE gear for our EMS workers. It has taken a while to get it back. In that time, we could have had a major catastrophe in the city—where would we be on the PPE gear, that kind of thing.

Chief PLAUGHER. The basic aspect of the Fire Act that is, we think, key, is the fact that it is direct to the local governments. And I understand the city's perspective, and obviously the match is of concern to all of us.

But it is a two-tiered match. It is based upon population, but it shortchanges all of the bureaucracy of the States that was alluded to during this session, and that is, that every time you add a layer, you add not only delays, but also groups that take slices of the money away from the intended purpose, whereas this is pretty straightforward.

Congress sets, in this particular year, $750 million. It goes to the U.S. Fire Administration. They have a small administrative overhead portion of it. The rest, a group of fire chiefs and other fire officials sit down and they go through each grant in its own separate category, and then they make awards; and then the awards come to the city through a notice, and you go about doing what you need to do to accomplish what the grant set out to do.

So it is not cumbersome, it is straightforward. Just exactly the opposite of that are the terrorist grants or the homeland security grants. They are, first off, almost impossible to figure out where they are and how you get them. And, second, they are extraordinarily cumbersome. You know, you have to do a whole host of things to prove that, a, you are buying an item off of an approved list, and, b, that it is something that fits into a program yet to be determined. And so it is—you know, there are so many parts that are very, very vague and difficult and hard to administer.

I know in the Commonwealth of Virginia which—we have been very fortunate, and nobody is criticizing Congress for how the Commonwealth of Virginia is being treated. We have received $204 million of homeland security grants, and that is a large amount of money. And, you know, very large State, very large population.

However, out of the $204 million, and I am talking about out of grand total sum of $283 million——

Mr. SHAYS. What is the population of Virginia?

Chief PLAUGHER. About 7 million. Arlington County, out of that $204 million, we are going to eventually receive about $600,000. And we were talking about before, about the threat, the nature of threats and obvious concern for certain items.

You would think that our needs would be addressed first because of the nature of what we have in our area. But that is not how the program is run. Strictly by population, each community gets a sum of money, and then they add a certain amount per capita on top of it.

So it is turning out to be a very difficult thing for us to administer and to provide what we think is necessary.

Mr. TIERNEY. Thank you. You have put on the record for us exactly what I thought you were going to put on, but I think it is important that we do that.

My fire chiefs have made this extremely clear. At my request, they have forwarded communications to Secretary Ridge and the White House. I suspect that the fire units across the country are doing that. I thank you for that. And maybe we will get some results there.

Chief Thompson, just let me wind up by saying that we have made a couple of attempts during the appropriations process to see if this administration would be willing to waive the match on a 1-year or 2-year term, whatever; try to get some of these grants out, because I know of the tremendous amount of expenditures on overtime that happened after the anthrax scare and the immediacy of September 11th. We have not had success so far, but we have not been that far away.

I think that we might be able to get some cooperative effort to keep trying to do that, at least to give you some breathing room between your problems that you have had because of the immediacy of it and after that.

So thank you very much.

Mr. SHAYS. Thank the gentleman. Mr. Ruppersberger.

Mr. RUPPERSBERGER. In the previous testimony that we talked about, the resources of money, and there's just not a lot of money there and we have to make sure that whatever we get we're going to make sure we get the best bang out of the buck, and there was an issue of regionalism. Now, you know, we have both the two departments connected I guess to each other, and it seems to me that if we could develop a formula where we could come and maybe apply for this money on a regional basis so there wouldn't be duplication of effort and especially in training, I mean there's certain moneys that you need. Equipment. But even with equipment if you all decide and there is a standard of the type of equipment that you need, maybe for biological and chemical as an example, what—how do you think that would work in the Washington Metropolitan Area. Is there that cooperation now? Does it need to be better? You know, putting together a program that the entire region asks for. And I think we'd all have a lot better chance. A region, wherever we are in the country to pull the regions together to get the resources that we need. Can you respond to those, both?

Chief THOMPSON. Actually working with Mr. Mike Burns, the National Capital Regional Director for Homeland Security, he's met with our COG Group, Council of Government Group in the jurisdiction area quite often in terms of our needs and what I think our necessities are for operating in case of an incident. He's been very supportive of what we're trying to do in terms of regionalizing our efforts in terms of equipment standardization and appropriation of funds to get the needed equipment, those kind of things. In addition, working with Chief Plaugher through the COG chiefs, we all sit down often at meetings and discuss the regional issues of operations and equipment. Communications gear being one of the primary things, being able to communicate at all with other jurisdictions in an incident of any magnitude or size. As evidenced by the Pentagon incident where if you have radios and don't have interoperability of them you have trouble communicating with the incident commander other than the old face to face or relay type, tag team type of thing. And we also coordinate efforts to make sure thats smoothed out, everybody's on the same frequency and same channel with the same type of radio to operate properly to stay safe.

Chief PLAUGHER. I thought the previous testimony was absolutely right on target about a couple of things, one of which is the standards. I think it's absolutely critical that we establish a set of national standards by which departments can judge their preparedness level and then it actually should be a very open process where you communicate back to your citizenry exactly where you are on this preparedness scale and the standards should be well established, well researched and they also alluded about it should be done quickly. We shouldn't now 2 years later still be trying to figure out what our standards are going to be in this arena. The fire service services are standard driven. We do almost everything by a set of standards and we have found a long time ago that 's the best way to protect our communities across this country.

The earlier testimony also talked about regionalism and about the need for that. One of the things that they did not however talk about during regionalism, and I fully support what they said about regionalism, was the fact that regionalism is how you leverage sustainability and long term success. If I am in a regional partnership with the chief here, there is no way that I'm going to let my relationship with him slip. So I'm going to guard that regional relationship with everything that I have and communities will do that, where our pressures will push and pull if it's a single community. If it's a single focused or single governmental entity other forces will push, pull—you were a county administrator, you know how that occurs. Push, pull, and the choices are then made.

However, regional efforts don't seem to take that same kind of beating. They seem to be sustainable. You can also then leverage through a regional effort where otherwise you cannot get. There is no need for every community or every fire department to have one or two or three of the widgets. Make sure that there's enough of the items that are around, enough trained people and that the region then has the capability to adequately respond within a reasonable and necessary timeframe. So I think the earlier testimony was right on target. And so I fully support the whole regional effort.

Mr. RUPPERSBERGER. Here's—but you need to talk about implementation. Because when September 11 hit our Governor pulled all the county executives of the seven major counties together and asked for a list of inventory that we needed so we could make a request. Well, some of us were right on and we came with what we thought was what we needed. Some others came with five times as much. Just that really hurt the process. So my—what I'm going to ask is that first you have county executives and the mayors that you have to deal with, it seems to me, and ask if you would consider going back to your first, your fire regional group that you meet on a regular basis and talk about the testimony today and let's talk about implementation and the possibility of coming up with some type of plan where you could agree maybe on the same type of communication equipment, maybe the same type of training, you know, whatever that may be. And I would hope we could do that with the police departments. Now, again, if it's going to work the boss has to be—the top person has to be for it. But if you could pull together and say this makes sense, that would help us, at least this committee with the leadership of Chairman Shays, to move forward and to try to do something about what we're talking about here today.

Do you think there's a possibility you could do that?

Chief PLAUGHER. Through our Regional Council of Governments those efforts are currently underway.

Mr. RUPPERSBERGER. I'm aware of that. And then you have Montgomery, you have Prince George's. I'm aware of that. But you know you get too bureaucratic sometimes if you get a lot of people involved.

Chief PLAUGHER. We're also very fortunate in the fact that we do have a coordinator. Mike Burn is our coordinator for the National Capitol Area and a lot of these regional efforts are being facilitated by his effort as well as the funding that is available.

Mr. RUPPERSBERGER. But would you all consider taking this back to your groups?

Chief PLAUGHER. Absolutely.

Mr. RUPPERSBERGER. As you're coming up with some type of formula on what you could at least agree to in the region and maybe the region beyond just Washington, Virginia and Maryland because you do have a national connection of governments, that might go a long way in setting a model or at least help getting forward to get what we need, because there's just never going to be enough money.

Chief PLAUGHER. Absolutely.

Mr. SHAYS. I thank the gentleman. In determining what the threat is and how we allocate resources, we would look at standards versus flexibility and we would look at all hazards versus threats specific. And one of the reasons you all like the fire grant is it basically is pretty general in nature and you can apply it where you want. But if we were basically to determine the threat assessment and we were going to allocate resources based on that there would be very specific things we would want you to get and other things we wouldn't allow to be part of that grant. Is that basically logical and acceptable or have we spoiled you in this general, you know, grant that you can apply to anything?

Chief PLAUGHER. Well, we also think there needs to be a baseline of capability in every community and so that's our starting point and that's what the Fire Act is so good at. Because there's a limited funding within the Fire Act, there hasn't been enough to move it past its basic categories that it's now funding and so that is why they're still buying—it was talked early about buying a lot of breathing apparatus, protective clothing, and also one of your committee members was talking about if departments had known that and they applied for it, they got the money, and that's still because we're just trying to cover the base front. We haven't moved past that. The needs were huge.

The first year of the Fire Act there were 19,000 applications and billions of dollars worth of requests and that's because the baseline piece hasn't been done. You have to then overarch that with preparedness for terrorism, and that's why this council's report, Council on Foreign Relations report says that the number is so huge is because we first have to cover the baseline and then we have to on top of that apply our ability to respond to a terrorist attack.

Mr. SHAYS. Let me just followup with you then. I mean my reaction is if—what should be Connecticut's responsibility for needs in Virginia or what should Virginia's responsibility be for needs in Connecticut? I basically conclude that it should be for the terrorist aspect of it. But I'm not sure it should be because you need a certain equipment that you should have anyway. That's kind of what I'm wrestling with.

Chief PLAUGHER. Yeah, the economic reality, however, is that our State and local governments are facing very, very tough economic times. And if we're going to provide protection for terrorism across the citizenry of our country, we've got to do a few basic things. And so I think that it would be an—in an ideal world, Mr. Chairman, it would be an ideal world to be able to take your approach and say every local government has to pony up for X number of baseline equipment and capability. But that's just simply not happening. So we can't get to the other end where we need to be, which is full preparedness, without that. So I think the assistance program, that's why it's called the Fire Fighter Assistance Act. It is there to just kind of nudge communities along and try to get them to focus on this issue. I think it needs to be in concert, however, with our standards making process, as well as our preparedness levels. Right now they're kind of in two separate tracks and they're kind of separated.

Mr. SHAYS. We asked both of you because you are right in the center of the storm. I'd like to know how well prepared you feel right now for a terrorist attack, be it chemical, biological or serious conventional or radioactive materials. I mean, tell me, in your mindset, do you think you're 20 percent of the way there, 80 percent of the way there or is it even impossible for you to respond to? Chief Thompson.

Chief THOMPSON. I'll speak very briefly. First, in all issues in terms of city preparedness, we're better prepared than we were 2 years ago, especially in terms of our first responder ability for EMS and fire suppression or HAZMAT mitigation or for biochemical attacks. Before we had access to funding we were woefully inadequate in terms of training and equipment. We have come to a certain level now. As the chief said, the baseline. We've got the baseline now. We've just got to maintain where we are and be prepared to go beyond that in terms of training and equipment issues that come about, new training issues that come about.

Are we ever fully prepared? Never that, because you can't prepare for any contingency. I mean, things happen that you don't expect but we're better prepared than we were previously. I'll let Chief Sellitto speak more to the issue of preparedness from his side, special operations side.

Mr. SELLITTO. Again, when you say you know how prepared are we, are we ready, you know, again, we don't really have something to measure it to. We've made leaps and bounds since September 2001 through the use of the Federal funding that we were lucky enough to receive quickly. We got the money. We were able to use it. In all areas, the training, the apparatus, the equipment, we're a lot further ahead than we were on September 11.

If you look at it, like I said, where do we have to be and what's the local responsibility compared to the Federal, I'm just going to give some rough numbers. We have roughly 1,200 firefighters, with roughly one quarter of them on duty at any given time. So we have about 300. We had about 350 masks, breathing apparatus on September 11, which on any given day is more than we need to operate, OK? But now you take a scenario like September 11 where we recall hundreds of men and maybe lost hundreds of those units in an event, is 350 enough? No. So we've gone out and purchased another 200 and again we got funding to do that. So we're sitting a lot higher than we were. Is that enough to cover every event? I can't say.

Mr. SHAYS. And they're trained to use that equipment?

Mr. SELLITTO. And we're trained to use it. And again we had over $4 million just for training and again that was——

Mr. SHAYS. And you've had a number of table top exercises?

Mr. SELLITTO. Numerous.

Mr. SHAYS. Chief.

Chief PLAUGHER. I think we're as prepared as any Fire Department in the United States, and I am talking about the Arlington County Fire Department, because of the work that we've been doing since 1995 for chemical attacks. So in that one arena I feel like we've made tremendous progress since 1995. The other areas of preparedness, and I'm talking about the full family of first responder preparedness, our colleagues in law enforcement, our colleagues at the hospitals, public health officials, those sort of things, they are woefully inadequately prepared for these and we're in the process of now of applying the few funds that we were able to receive from Homeland Security to that arena, buying protective clothing for police officers, hospital employees and that sort of thing. I think that when it comes to other key areas such as was addressed earlier this morning about hospital preparedness and hospital surge capacity, we haven't even begun to prepare for that. And we are, again, in sad, sad shape because again, the nature of the problem, as Dr. Smithson was talking about, is that it's such a difficult problem with the private sector and nonprofit piece of the hospital pie that's there as to preparedness.

So, again, to answer your question, Mr. Chairman, it depends upon preparedness for what arena. And so it goes across the entire scale of maybe the seven to eight for chemical preparedness and there's still room for improvement and resources can be applied all the way down to a negative something when it comes to hospital preparedness because we've not done well.

Regionally, we're nowhere near where we need to be. We're trying to embark upon a major effort in the Commonwealth of Virginia through the State police divisions. I'm forced to work through my Commonwealth of Virginia. The laws of my State say that I have to apply time and energy working through relationships within the Commonwealth of Virginia. That's where our funnel is for request for equipment and the systems that we have. And that's not a bad thing. That's a good thing to have a State relationship.

So again, we're preparing in a multiple of fronts. We're preparing in the regional, COG regional effort with our partners in both the District of Columbia and Maryland suburbs as well as we're preparing within the State arena. So there's a lot to do. But the regional pieces are nowhere near where they need to be.

Mr. SHAYS. I realize we just have firemen here, but, and I mean just in the terms of my question. Dr. Smithson was very clear that—and I spend nights with—in fire houses, which is a lot of fun for me. For one, I've never gone to a better restaurant. But two, just great people to talk with. But there is a lot of opportunity to train because you have—when you're not putting out a fire and so on and so, it's something I hadn't really focused on but you just tend to think in terms of training a little, I think more clearly. But we really did not have a lot of police departments say they wanted to come to this hearing. Is that because their basic first response is going to be crowd control and they feel they know that anyway? Or, I mean in your dialog with police chiefs, how do you guys view your various roles? I mean——

Chief THOMPSON. Well, in terms of city function with MPD and how we coordinate and liaison together, pretty much they're the law enforcement end of security, securing a site, protecting a site and making sure that we have access to a site. They also have the capability, MPD does now, of if an incident occurs and it's of a criminal nature; they have equipment to go in and make the investigation possible by special equipment they have purchased recently. We work very well together with them on incident command systems in terms of a large incident or a small incident, how we coordinate our efforts together and definition of roles and who's got responsibility for what.

From my standpoint with Chief Ramsey and his group, MPD side, they understand their role is primarily to secure the site if a law issue is there, MPP would be there to handle the law end of it without any suppression mitigation of the incident.

Mr. SHAYS. Obviously if they went to a site they might need equipment to protect themselves from chemical exposure and that type thing?

Chief THOMPSON. Based upon if they arrive on the scene of the incident first without any prior knowledge, quite naturally they'll call us. But then again if they come on the scene, we're there, they got pre-knowledge of equipment they may need to bring. They have the equipment that they have purchased for their use, yeah.

Chief PLAUGHER. We work daily with our colleagues in law enforcement, both with the police department and our sheriff's department, are two key elements within our ability to respond as an emergency team. They are in the process of being better protected. Each one of our law enforcement officials has been trained for basic knowledge and has some protective equipment. But they have a huge role to play in maintaining the civil order of our community in the event of any terrorist attack and their visible presence as well as their knowledge and ability to respond to the citizenry's questions about what is going on and what actions the citizens should take or not take is absolutely vital. You don't want that blank look on a police officer when a citizen comes up and asks them what to do. They need to be able to articulate exactly what the plan is and how things are to be unfolding. In the event of a bio attack, I say that the law enforcement slice of our community will be vital and their ability to adequately respond to a bio attack is monumental. The need to provide for security of our medical facilities, the need to be able to adequately maintain civil order I think is high on the order, and that's going to mean protecting those law enforcement officials themselves as well as their participation within our incident command structure and systems that the chief was talking about earlier.

So I think it's unfortunate that they're not here. It's unfortunate that the International Association of Chiefs of Police aren't here to talk about this report because my estimation is that they would be saying that this report is also very much on target.

Mr. SHAYS. Thank you very much. Anything?

Mr. RUPPERSBERGER. Just one comment, Mr. Chairman, and the question you asked, and maybe I got it right or wrong, but the issue of whether or not the resources that would come would just be dealing with the issue of first responder terrorism and would that really be the focus of it. And it seems to me that everything that you're talking about, whether it's biological, chemical, we have HAZMATs, you know, things like that, that these resources you'll get could also be used in natural disasters. I mean the training, it's about a system. It's not about just given this. It's about the system you create that works, from the first responders, the paramedics, whether it's the suppression, the police coming together and then taking victims to hospitals and how we get to the hospitals, your communication systems.

So in the end, I mean I have always believed if one door shuts another opens, you know. In the end if we get this together we will be better for this years to come, if we can ever get it together right now, and that's what we're talking about here today. Do you agree?

Chief THOMPSON. Absolutely agree. Sure.

Chief PLAUGHER. I think all hazards and the all hazard approach is absolutely vital to our preparedness. If you're not prepared for a hurricane you're not prepared for a terrorist incident. If you're not prepared for a terrorist incident you can't be prepared for a hurricane.

Mr. SHAYS. Right. The issue that triggers to me though is you should before September 11 have been prepared for some of these. And where does the Federal Government step in to make sure that we have a protected system around the country for this new revealed phenomena, which is terrorist attack; in other words, the human-induced crisis?

Chief PLAUGHER. Well, the Oklahoma City attack, which was obviously an early warning from my perspective, an early warning event for this country, should have moved us further along in the preparedness and I allude to earlier hearings that were held here and you know you were having a hard time getting people to even talk about the subject back in those days. And so I commend you for your efforts. You made a great deal of effort to try to move us along on that scale and we're now starting to see a diminishing of interests in this arena and we just simply cannot let that happen because we're not there. We are not there on the natural level and we're not there on the mandate level either.

Mr. SHAYS. I hear you, and we need to do that. What I'm wrestling with is I ask myself if this had been a Democratic administration, what would I as a chairman of a committee be asking for, and I think I would be asking for more than I have been asking, you know, which is a good wakeup call for me.

Mr. RUPPSERSBERGER. Well, I'm glad you said that because that's leadership.

One other comment I could make. The police aren't here, but I think it's important, another role that we haven't talked about, but if resources are going to go to our police departments there's a lot more resources that need to go into intelligence because if you look at what really in my opinion has deterred another September 11 incident, is the—not only the intelligence but the cooperation between the CIA, the FBI, the NSA, the State and locals, including educating the public because a lot of leads will come from the street. So that is another area that from this committee point of view we're going to get resources where we have to put in money into this.

Chief PLAUGHER. And when you do that look real hard at the fusion center concept because they are absolutely critical, that we build fusion centers and we build a regional team of fusion centers so that we can maximize any and all intel into a workable product and that it goes across the entire spectrum of the response community. In other words, that it has an opportunity to feed all fire EMS as well as law enforcement communities and that nobody gets missed in the product of a fusion center. And I think they're going to be vital. I don't see how we can do it without it.

Mr. SHAYS. This is where you use the classified and the every day information and put it all together?

Chief PLAUGHER. Yes, sir.

Mr. SHAYS. Yes, I have the view that had we done that we would have probably known about September 11.

Chief PLAUGHER. Yes, sir.

Mr. SHAYS. Thank you all very much. Any last thing on the comment before we adjourn? You have been a wonderful panel, and thank you.

Chief THOMPSON. Just like to say one thing, particularly in terms of issues we've discussed here in terms of standardization of training and equipment across the country. We interact with the COG group as a COG group with the other jurisdictions in terms of equipment and purchase of equipment and how they'd be compatible for use. But if Chief Plaugher's group goes down to somewhere in Loudoun or in Richmond and the equipment is not so much as standardized on a national level, they operate in a vacuum because they can't work together. They just can't do it. It's almost impossible, to work together. Then you have a catastrophe to respond to an incident they can't be prepared for, they're not prepared for. You need national standardization of training and equipment. It's as simple as that.

Mr. RUPPERSBERGER. And communications.

Chief THOMPSON. And communication, absolutely right.

Mr. SHAYS. Thank you very much. This hearing is adjourned.

[Whereupon, at 12:26 p.m., the subcommittee was adjourned.]

○

APEX
TP
*9780160715662*
76901945 – 98